Guide to Effective Grant Writing

Guide to Effective Grant Writing

How to Write an Effective NIH Grant Application

Otto O. Yang

Departments of Medicine, and Microbiology, Immunology, and Molecular Genetics
Geffen School of Medicine, UCLA Medical Center, Los Angeles, California

 Springer

Library of Congress Cataloging-in-Publication Data
Yang, Otto O.
    Guide to effective grant writing : how to write a successful NIH grant application /
by Otto O. Yang.
       p.   cm.
    Includes bibliographical references and index.
    ISBN 0-306-48664-4 (pbk.) – ISBN 0-306-48665-2 (ebook)
    1. National Institutes of Health (U. S.) – Research grants – Handbooks, manuals, etc.
    2. Proposal writing for grants – United States – Handbooks, manuals, etc. I. Title
    RA11. D6Y36 2004

ISBN 978-0-306-48664-7   2004051680
ISBN 0-306-48664-4 (pbk.)
ISBN 0-306-48665-2 (eBook)

Printed on acid-free paper.

Printed in the United States of America.

10 9 8 7 6 5 4 3

springer.com

This book is dedicated foremost to my parents, who have provided constant support for all my endeavors. It is further dedicated to all my teachers, who provided me the tools to think and write. Finally, it is dedicated to my wife Wendy, without whose patience and understanding my work would not be possible.

Foreword

The bane of the academic existence is grant writing, and yet without grants most of us would not be able to continue to ply our trade. With this enormous importance of grants, it is surprising that we are so well trained to do research, with almost no formal attention given to grant writing, which is the lifeblood of academic research.

Help is now at hand—this book by Otto Yang MD is the first truly helpful guide to grant writing. The book is likely to be of great help to anyone writing a grant, even those of us who are more seasoned grant writers. Organized in similar fashion to an actual NIH grant, with specific sections on specific aims, background and rationale, preliminary results, and experimental design, it outlines in very readable fashion specific suggestions for convincing study sections that the ideas being presented are worthy of funding, and turning the reviewer into an advocate for the project being proposed. It also points out common errors that make reviewers lose enthusiasm even when the experiments are highly worthy of funding. Frequent use of examples makes the points very clear, and the clear style makes the book an enjoyable read.

As a person who is often asked to serve on study sections to review grants, I owe a debt of gratitude to Dr. Yang, since anyone using this book will undoubtedly deliver a grant that is much more accessible to the reviewers. Although there is no substitute for great ideas, all too often these are hard to dissect from the disorganized and poorly presented grants. For those who follow the suggestions in this book, the outcome is certain to be improved.

BRUCE D. WALKER
Boston MA 4/27/04

(Dr. Walker is a Professor of Medicine and the director of the Partner's AIDS Research Center at Harvard Medical School. He is also a former chairman of an NIAID/NIH study section, and recipient of the Doris Duke Distinguished Clinical Scientist Award.)

Preface

A crucial skill as an academic faculty member in the health sciences is the ability to get research funding. Doctoral and postdoctoral training are highly focused on acquiring the scientific tools required to pursue a career in research, but learning the complementary skills of building and managing a research operation is usually left to observation, trial and error, and occasional elective seminars. In my own career as a physician-scientist, I've been extremely fortunate to have outstanding role models and mentors who have walked me through the process of grant writing and helped me navigate the process of submitting grant applications to the National Institutes of Health (NIH). Without such guidance, trial and error would have been an inefficient and perhaps disastrous method to learn these skills.

Serving on an NIH "study section" (IRG) that reviews grant applications in HIV pathogenesis, I've now had the opportunity both to assess hundreds of grants and to see the process from the perspective of the reviewer. The central importance of writing effectively has become especially obvious after being in this role. It has been surprising to what degree certain mistakes and problems recur, and how much impact the presentation of a grant affects its reception during review. It struck me that many of these errors are simply due to lack of experience or formal guidance, and that many grants could be significantly improved with just a few pointers. This guide to grant writing is therefore intended to fill a gap in the formal instruction of academic faculty in the skills of presenting a research project to NIH for funding, a chief mechanism of support for many academic researchers.

The information and advice provided here are intended to assist in the process of applying for NIH grants, which are a major funding staple for developing and established scientists in the health sciences. However, this guide absolutely is *not* a substitute for the detailed step-by-step instructions and advice provided by NIH. Those resources are essential to the mechanics of producing and submitting a grant, and should be followed strictly. The focus of this guide, however, is writing to present scientific ideas clearly and avoid common mistakes that detract from the application. The principles given here are based on my own opinions, which have been shaped by numerous helpful discussions and critiques from my mentors and colleagues, and my reading of applications as an NIH grant reviewer.

In the end, the strength of the scientific merits of any application in the eyes of the reviewers will determine its score and ranking. Far and above any other consideration is the quality of the ideas and research plans; a weak project, no matter how well presented, will not score well. Obviously, no brief guide can provide instructions on how to be an outstanding scientist. However, even a very innovative and promising project may be presented in a manner such that the reviewers fail to appreciate its quality and promise. The presentation of the grant is, justifiably or not, perceived as a reflection of the thought process and carefulness of the investigator, and therefore carries significant weight. Given the large number of applications received and the intense competition for funding, the quality of writing, sometimes referred to as "grantsmanship," is commonly a factor that tips the scale between acceptance and rejection.

This guide is by no means a definitive presentation on the ideal application, as there is no single "best" method to construct an NIH grant application. Obviously, there are no tricks or strategies to guarantee success. The style of presentation depends greatly on the subject matter. However, there are certain characteristics shared by almost all well-written grants, and recurrent themes common to many poorly written grants. The following chapters are intended to illuminate some of these concepts.

Finally, while the advice given here pertains specifically to NIH grant applications, the principles are applicable to applications to other institutions, such as the Doris Duke Foundation, National Science Foundation, American Cancer Society, and numerous others that provide research funding. The structures of most grant applications are usually very similar or even identical to that of NIH, because the same questions need to be answered regarding the rationale and plans for your work, no matter which institution.

I hope that this guide will assist you the reader in presenting ideas clearly to grant reviewers, to ensure that your applications will be seen in the most positive light possible. I welcome any comments, criticisms, and corrections, because undoubtedly this guide, much like most grant applications, will have ample room for revisions and improvement.

Contents

1

Overview: Overall Goals When Writing Grant Applications

Getting (and keeping) the attention of the reviewers

A research laboratory can be considered analogous to a business in many ways. Like a business, there is management of workers and finances to make products (data). Also like a business, income must be earned in a competitive process. Grant applications are the presentation of your work to persons who decide whether they want to invest and buy, much like the role of product advertisements for a business. Clearly, having products that are so superior or important that they sell themselves is ideal, but most are usually in competition with other similar products. The manner in which your products are presented to potential investors is absolutely crucial, to convince them to choose yours over others. Much like for a harried consumer, the attention of a reviewer needs to be quickly captured and sustained, to "sell" your ideas effectively.

Presenting your research ideas in a grant application therefore shares some principles with advertising. A major goal is to convey the major points in a very efficient, succinct, and interesting manner. Grant applicants often assume that the reviewers will comb through their application in great depth, but this is somewhat analogous to expecting a consumer to take detailed notes on a magazine or television advertisement. Of course the grant reviewers *will* spend time and energy trying to absorb the presented material, but minimizing the required effort to do so (and making it as easy as possible for the reviewers) is important, just as in advertising. The more interesting and less laborious the process of reading the application, the more likely that it will be received positively.

1

Balancing between clarity and depth

Unlike commercial advertising, however, the target audience for a grant application is highly knowledgeable about the area you are discussing. The reviewers will be experts in related areas of research, and the depth of your thinking and technical capabilities will need to be conveyed. This means that there is a fine balance between being succinct and being thorough in the material you present. The extremes are presenting the ideas in broad strokes versus presenting a comprehensive list of experimental protocols. Deciding where to set the balance between these extremes falls is one of the more challenging aspects of grant writing, because this varies depending on the specific project and context. However, there are a few general principles that can assist in finding this balance.

Careful organization of your application can increase the information you can include without losing clarity. For example, making the "Specific Aims" section a very clear summary of your ideas without technical details, and confining those details to highly organized sections directly pertinent to each aim in the "Research Design and Methods" section can allow you to present a large volume of details without clutter. This is analogous to describing a piece of machinery with many parts: if all the parts are depicted in assembled form, it is easy for the reader to see how the machine will work, but if the parts are simply listed in no particular order, along with a description of how they should fit together, it is far more difficult and tedious to understand.

Being strategic in choosing the details to include or exclude is another method to simplify the balancing of clarity and detail. The ultimate goal of your experimental descriptions is to convince the reviewers that the work is feasible in your hands. Your explanations should be prioritized to meet that goal, rather than being completely exhaustive. The emphasis and detail should be just enough to show that you have put full thought into the issues. For example, you may describe a new or difficult technique in great detail, documenting the sources of all reagents and potential technical pitfalls, while you may describe a common technique in broad outlines. This approach tailors the material you include, shows the reviewers you are thinking and prioritizing, and avoids unnecessarily burdening (and boring) the reviewers with unneeded details.

Telling an interesting story: the art behind writing the application

Research progress is very much like an ongoing story, with plot twists and surprises. A well-written application creates a tale that appeals to the reader. The plot is revealed in the "Background and Significance" section, laying out a self-contained story. Unlike a novel, however, the story is unfinished. After the plot is presented, the reviewers should be curious about what happens next, and the questions you propose (the specific aims) should reveal how you will unfold the next chapter. By the time the reader has finished reading the "Background and Significance" section, the full context for your questions should be clear, and the issues addressed by the aims should already be on his or her mind.

This analogy can be instructive for how you will arrange the information. If you have multiple lines of investigation, it is crucial to tie them together, like multiple plots in a story. Like a well-written story, all the pieces should fit together, without extraneous information. A common mistake is the tendency to use the "Background and Significance" section to showcase comprehensive and exhaustive knowledge, without a clear thread of thought. This rarely impresses a reviewer, and demonstrating how you think and prioritize is usually more important than displaying your breadth of knowledge in your research area. To continue the analogy, it is more interesting to read a well-written novel than a volume of an encyclopedia.

2

Organization and Use of This Guide

This guide is organized in chapters that address specific aspects of grant writing or individual sections of applications. Each chapter covers concepts that are important to writing an effective application, and explores how to approach them as a writer. Where relevant, the chapters end with a list of common errors, and discussions of how to avoid these pitfalls.

Chapter 3 describes issues that need to be considered before a grant application can be submitted to NIH, such as administrative deadlines at your institution.

Chapter 4 introduces the standard format for all NIH grant applications. Explanation is given as to why strictly adhering to this scheme in your applications is crucial.

Chapter 5 provides an overview of NIH grant programs, with a brief explanation of some common types of grants and their goals.

Chapter 6 gives an overview of important technical principles in writing an application that is as clear and attractive as possible for the reviewers.

Chapters 7 to 11 cover the individual sections of the scientific portion of grant applications: "Specific Aims," "Background and Significance," "Preliminary Results," "Research Design and Methods," and "Literature Cited." The purpose and organization of each section is described, and suggestions are given on how to write these portions of your application.

Chapter 12 describes the proper use of appendices to bolster your application with supplementary information.

Chapter 13 reviews the required content in the administrative sections of grant applications. The importance of being attentive to these sections is emphasized and explained.

Chapter 14 discusses the involvement of collaborators and consultants to strengthen your grant application.

Chapter 15 explains the process of grant scoring at NIH, giving a perspective on how each application is evaluated by reviewers, and how various criteria contribute to the final score and funding decision.

Chapter 16 explores strategies for responding to reviewer critiques for resubmission of applications, with advice about how to interpret reviewer comments, and how to respond to the criticisms with revisions of the application.

Chapter 17 discusses issues and strategies specific to writing a competing renewal application for a previously funded grant.

Chapter 18 discusses the general principles of writing applications for grants by non-NIH institutions.

3

Preparing to Write

NIH provides a wealth of information concerning all aspects of grant applications, including the different types of grants available, research priorities for funding, the mechanics of writing a grant, deadlines, contact information for advice and guidance, and explanation of the review process and funding decision. A great place to start is the Internet, where application forms, instructions, deadlines, study section listings, and other aspects of the application process are available (see Appendix).

There are several considerations before you start writing your grant application. Although your main concern may be the scientific portion, there are many other factors that affect the success of your application and thus demand your attention. These include:

Deciding the grant for which you want to apply

Two major factors determine the grant for which you should apply (see Chapter 4). First is the stage of your career, which determines whether a training grant, career development award, or independent investigator research grant is appropriate. Second is the stage of your research. If your project is highly preliminary and exploratory, a grant that targets innovation without requiring preliminary data (e.g., R03 or R21) is a good match. If your project is highly developed with ample preliminary data, a larger research grant (e.g., R01) is your likely goal. When unsure, you should seek advice from mentors or colleagues, and speak to an NIH program officer for guidance. Matching your project to the purpose (e.g., training a new investigator versus producing important scientific results) of the specific grant is crucial.

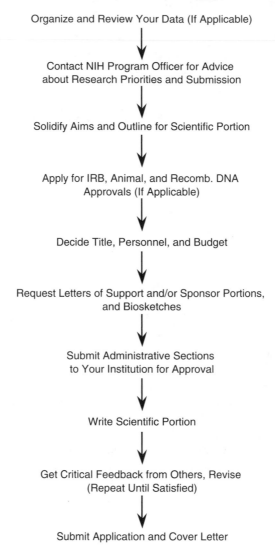

Organize and Review Your Data (If Applicable)

Contact NIH Program Officer for Advice
about Research Priorities and Submission

Solidify Aims and Outline for Scientific Portion

Apply for IRB, Animal, and Recomb. DNA
Approvals (If Applicable)

Decide Title, Personnel, and Budget

Request Letters of Support and/or Sponsor Portions,
and Biosketches

Submit Administrative Sections
to Your Institution for Approval

Write Scientific Portion

Get Critical Feedback from Others, Revise
(Repeat Until Satisfied)

Submit Application and Cover Letter

Determining whether your research
falls in an area targeted by NIH

If the topic of your work falls into an area targeted for funding by NIH in
a "Request for Applications" (RFA) or "Program Announcement" (PA), you
may improve your chance for funding if you submit a response instead of an
"unsolicited" application. NIH uses these mechanisms to encourage research in
particular areas or topics, and provides detailed instructions on the objectives.

When choosing to respond, you should review the RFA or PA thoroughly, because your application can be rejected without review if the responsible program officer deems that your project is not applicable. Discussing your project with the NIH program officer issuing the RFA or PA before applying is highly advisable.

Timing of Deadlines

Several deadlines other than the NIH submission deadline (three cycles a year) are important. Your institution will require time for an administrative review before the application can be submitted to NIH. This process can be started well in advance of submission. Usually, your institution will only require the face page, the description/performance sites/personnel page, the budget page, and the checklist for this approval (see Chapter 13). If you plan ahead, these require little effort and will save you last minute headaches. Thus you can give these documents to your institutional administrators for approval while you are writing the scientific portion of the application.

Also keep in mind that there are several months between submission and the earliest date that funding could start, even if the application is accepted. If you need to plan your research finances, this is a key point.

Institutional Approvals

If your research involves human subjects, vertebrate animals, or recombinant DNA, institutional approvals will be required. Even if the work uses samples from human subjects but is IRB-exempt, documentation from your institution will be needed for your application. NIH will not accept your application for review unless these approvals are in place or under review at your institution (see also Chapter 13 on the administrative sections).

Application Strategy

Funding of your work will depend on the priorities of the reviewers who read it. These priorities will depend on the NIH institute to which you submit the application, and on the specific study section ("integrated review group" or IRG) within that institute. A project may have a focus or slant that makes it more interesting to one institute and/or section than another. You should also check to see what similar types of projects have been funded at each institute, and what research priorities have been stated in issued "Requests for Applications." Doing these background checks can help ensure that your application reaches the reviewers most likely to appreciate your work.

You should contact a program officer at the institute to which you are considering submission, for advice on the appropriateness and likely interest in your work. It is also useful to get advice on what study section (IRG) would be the most appropriate to review your application. NIH provides listings of recent members on each study section, and you may want to review these lists to determine the section that would be best for your application, as well as to identify members whom you would wish to exclude from the review. In most cases, a request to send your application to a specific study section will be honored. See also Chapter 13 on the cover letter, and Chapter 15 for the exclusion of reviewers.

Estimating the Budget

You will need to have an idea of the resources you will require for the project. Assuming that your grant request will be less than $250,000 per year, your budget will be "modular" (see Chapter 13) and you will not be required to submit a detailed accounting except for salary costs. However, you should still have a good idea of how much money you will actually need for the work, and how to justify the amount. For larger grants ($500,000 per year or more), preapproval from NIH is usually required before submission of an application.

Collaborators and consultants

If you are not a well-established independent investigator, it will be crucial to demonstrate that you have the necessary guidance to be successful. For training grants, this means having a mentor who will show firm commitment and elaborate a clear training plan. For research grants, this means having collaborators and/or consultants who will ensure that you have all the technical assistance and reagents necessary for your project. In either case, make sure you allow plenty of time for letters of collaboration or mentor's sections for your application.

Having the support of an acknowledged expert can be a boon to your cause. It may be useful to note the members of the study section who will review your grant; persons who are familiar with the work of you and/or your collaborators are likely to have a more optimistic impression of your abilities. Be aware, however, that the membership of the review group is subject to change and often shifts. Also, if you are applying as a primary investigator, it is important to ensure that your independence as the lead investigator is not compromised by the role of collaborators or consultants.

Familiarization with the NIH format

Writing the application in the proper format is crucial, as is discussed in detail in this guide. Be thoroughly familiar with the format required by NIH, as this will make your application as easy to read as possible for the reviewers and administrators. See also Chapters 5 and 6 for an overview.

Obtaining NIH forms for the application

NIH provides all the necessary forms and instructions on their website. These are available online in PDF (Adobe Acrobat) or RTF (rich text file, compatible with most word processing software) formats (see Appendix).

4

Types of NIH Grants

The list of types of grants is long and covers the gamut from small awards to encourage and train new researchers, to huge grants covering groups of senior scientists in collaborative projects. These grants are divided into multiple series. All have in common, however, the same presentation format. Each series has a general theme, and individual grants in the series have a focused goal, whether to support postdoctoral training projects or to fund established researchers doing advanced research.

"F" awards are "Individual Fellowship" awards such as the F32, intended to support postdoctoral fellows in preparation for a career in biomedical research. These are used to support training with the goal of future independent investigation, under the mentorship of a senior investigator.

"K" awards are "Career Development" awards, intended to develop the research careers of investigators who are in training to become independent. These include grants such as the K08 "Clinical Investigator Award," to support physicians who wish to develop careers as scientific researchers, and K01/K02 "Research Scientist Development Awards," to support junior research scientists in their training.

"R" awards are "Research Project" awards, funding targeted research investigation by independent investigators. Examples include the R21 "Exploratory/ Development Grant," which is focused on innovative ideas, and the R03 "Small Grant" to explore and develop high-risk ideas. Until recently, R21 and R03 grant applications could only be submitted in response to an RFA or PA, but NIH now accepts unsolicited applications of these types. Also in this group is the R01 "Research Project" that is the standard staple of support for unsolicited or solicited research projects by independent faculty; this is the major NIH biomedical research grant on which many academic researchers depend.

Targeting the content and emphasis of your writing to the type of grant

These grant types each have different goals, page limitations, budgets, and research priorities that affect how they are reviewed. You should thoroughly familiarize yourself with the criteria for the grants to which you apply. A full understanding of the NIH purpose behind the grant will assist you in writing it effectively, because this will allow you to focus your application on fulfilling this purpose. The following points should be checked and kept in mind when you apply for a specific grant:

1. *What is the goal of the grant from the standpoint of the NIH?* For F and R series grants intended to train research fellows, the emphasis will be on the training aspects, such as the training track record of the mentor, the educational plans described in the application, and the potential for the scientific project to provide useful experience to the applicant. For R series grants intended to support independent researchers and to produce important scientific advances, the emphasis will be on the track record of the applicant in prior independent research projects, the validity of the ideas and approaches, and the potential impact of the results that would be produced by the research.

2. *What research results are expected from the grant?* As above, F and K series grants are chiefly concerned with training the applicant, and the actual scientific results are secondary. Different R series grants are intended to produce important scientific results, but vary in emphasis. For example, the R21 favors innovative and high-risk exploratory projects with high potential impact, while the R01 favors projects that are more fully developed and sure to produce progress.

3. *What is the research focus of the grant?* This is especially applicable to grants given through a "Request for Applications" (RFA) or "Program Announcement" (PA), which must address a question or area of interest targeted by NIH in the RFA or PA. In this case, the manner in which you tie your work to the stated priorities of NIH is crucial. For "unsolicited" grants (not responding to an RFA or PA), the burden will be on you to explain and justify the importance of your project.

4. *What are the administrative requirements for the application?* The administrative staff at NIH rejects applications that do not meet the requirements for length, format, human subjects and animal documentation, etc.

Again, it is absolutely essential to understand the motivations behind the grant for which you are applying; NIH provides money to meet its own agendas and goals. There is no substitute for the instructions and information issued by NIH, and you should therefore review them carefully so that your application addresses the stated purpose of the grant program as closely as possible.

5

Anatomy of the NIH Grant Application

The purpose and importance of adhering to the standardized format

All NIH grant applications are presented in a highly standardized format that should be strictly followed. While intuitively it may seem that varying from this format might give your grant a more original flavor and help it stand out, in fact it is almost certain to detract from its appeal. The reason is that this rigid organization provides a standard framework that makes it easier for reviewers to read and assess multiple grants. Each reviewer must deal with several applications in detail at once. At the study section meeting to assign priority scores, each reviewer skims through dozens. When a grant is in the familiar standard format, it is simpler for the reviewer to locate specific information within the application quickly.

The primary and secondary reviewers who read the grant in detail provide the written reviews. They are themselves busy researchers, and they therefore read your grant application (and a stack of others) whenever they have free time. Because it can take hours to read an application in detail, this means that they usually read during multiple sittings. These reviewers thus often need to re-scan the text to remember things that they may have read days or weeks earlier, before sitting down to write a formal review, and again to present and discuss it at the grant review meeting. At the meeting, other reviewers seeking information about the project may want to find it by quickly scanning the text during the discussion.

For these reasons, it is crucial that all the information is easily locatable by a casual reader. This is accomplished far more readily if it is organized in the standard format.

It is therefore to your advantage to keep everything organized as tightly and clearly in this standard framework as possible, because easier access to the details of the project makes it much clearer in the minds of the reviewers. Deviating from this format makes the job of the reviewer more difficult, and is likely to leave a negative impression even if you feel your project would be better organized another way. Busy grant reviewers find spending extra time to find information in an unfamiliar format cumbersome and unappealing. So, while it may seem to you that your ideas could be more clearly presented by altering the NIH format, and that it is tedious and unnatural to adhere to such a rigid system, you almost certainly will find it disadvantageous to stray. In the end, it is the convenience of the reviewers and the NIH that are paramount. Also, you will find that it becomes quite simple to organize your thoughts in the NIH format, the more grants you write (and read).

Detailed instructions on how to write and submit an application are provided by NIH, and you should thoroughly familiarize yourself with this information (see Appendix for how to access forms and instructions online). The instructions provided with the application forms are your primary resource for the rules and regulations concerning your application. Helpful hints on the format of grant applications, the application process, the review process, and the funding process are available through the internet and NIH program officers. Information in this guide is purely *advice* to assist you in the process, and is not intended to supercede any of the information and guidance provided by NIH.

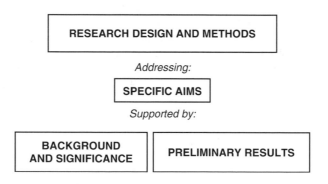

Structure and organization of the standard fomat

The scientific sections of the standard NIH grant format are organized under headings titled: *Specific Aims, Background and Significance, Preliminary Results/ Progress Report, Research Design and Methods*, and *Literature Cited*.

The *"Specific Aims"* section is a short and self-contained summary of the research questions and the goals of the project. It should stand alone and function both to give an extremely brief introduction to your research project and to jog the memory of someone who has read your grant earlier and needs a brief reminder of the project.

The *"Background and Significance"* section should thoroughly but concisely make the case for why this research is important, provide a context for your work, and indicate how results from the project would affect the field. Because the reviewer may be an expert in a related but separate field, this section should serve as a review article for a sophisticated reader and bring him or her up to date in your field. NIH suggests that the style and tone of the magazine *Scientific American* is a good standard; this background should tell an engaging story, as described in the previous chapter.

The *"Preliminary Results"* section is a demonstration of your ability to perform the proposed work, illustrating technical tools and experimental prowess to approach your aims. For some grants (such as R03 and R21 applications) this section is optional because preliminary data are not required, but it is a central component of the standard R01 grant. For competitive renewals of R01 and other renewable grants, the *"Progress Report"* is also included in this section to demonstrate the progress toward the aims of the prior funded grant.

The *"Research Design and Methods"* section is a detailed outline of the proposed research, and should serve as a blueprint for the work to be performed if the grant is funded. It should provide enough detail to demonstrate a clear direction for the work, and the technical capability to perform it. All experiments should be organized under the "Specific Aims" section, demonstrating how each aim will be accomplished.

The *"Literature Cited"* section is an inclusive bibliography of published references pertinent to your work. While often ignored as being secondary documentation, it is actually important in the presentation and can affect the evaluation of your application.

Page limits

This entire scientific portion (excluding the "Literature Cited" section) should fit within the dictated limits for the application, which varies by grant type. For example, the R01 is usually limited to 25 pages of single-spaced text. Observing these page limits is important, because NIH can return an application without review if it is too long. Because space is limited, layout, organization, and style are especially important determinants of how your message is conveyed. The following chapters cover each section of the scientific portion in detail, and discuss common shortcomings and pitfalls in the writing of these sections.

6

Organization and Aesthetics

Making a first impression

Organization and neatness have an impact on the reader. They do not factor into the official scoring criteria, but the reviewers' reactions are colored by the presentation of your application. As will be repeatedly stressed in later chapters, clarity and ease of reading are crucial to convey your ideas as positively as possible. Although the intrinsic value of your ideas is the final criterion of the success of your application, skillful presentation is crucial to convey the importance of your project as fully as possible. Moreover, characteristics of the document itself may have some effect, conscious or subconscious, on the reader. A document that is disorganized or displeasing to the eye is unpleasant and difficult to read, and may be interpreted as reflecting lack of care and effort in the planning of the project. It is thus best to be attentive to details concerning how your grant is presented, just as a gourmet chef may dote over the plate and garnish used to present a food creation.

Basic rules set by NIH

Some very basic rules on grant organization and aesthetics are set by NIH, and you should review these carefully before writing your application. Organization of information into several major sections is clearly dictated in these instructions. Also, basic criteria are specified regarding page margins, font sizes, and figures. The page margins are to be at least half an inch, the font is to be at least 12 point, and the figures are to be large enough so that all text and symbols are easily legible. Violation of these guidelines is grounds for rejection of your application without review, because these rules are designed for the benefit of the reviewers and NIH staff. Still, staying within the basic guidelines does not guarantee that the application is aesthetically acceptable, and it is easy to follow the rules to the letter while violating their spirit.

Text and figure organization

Generally, the document should be easy on the eye in terms of the quality of the text and figures. Organization and planning can make the difference between a haphazard or polished appearance of an application. Clarity of the layout can enhance and emphasize the clarity of your ideas, while disorganization can make it difficult for the reviewer to visualize the structure and thematic relationships of the proposed work, and render key information difficult to locate. Attention or inattention to detail shows quite clearly when the reviewers read your grant, and can affect their perception of how you approach your work in general.

Writing style and organization: following an outline structure

Although there are many effective styles of writing, all successful grant applications share the characteristic of presenting information in a clear outline format. Most successful writers begin with a mental if not literal outline before putting down the first word; this is extremely useful for both the writer and the reader. Organizing all the information in an outline structure makes the task of recording and assimilating the information much easier. For the writer, it is a way to previsualize the whole project, see how the parts will fit together, and foresee any logical or experimental holes. The entire grant can be considered as a unit before anything is written, and the logical train of thought established from "Background and Significance," to "Preliminary Results," to "Research Design and Methods." Predetermination of how all the parts fit together makes it simpler to write each section without the need for frequent adjustments and revisions for omitted or redundant points, or poor transitions. For the reader, an outline format provides fixed points of reference within a long and complicated document, making it easier to find key information and absorb the message. Because each cluster of facts or protocols supports a topic sentence, a review of each topic sentence quickly summarizes all these items in a logical framework. This facilitates understanding the point behind each item, and the relevance of each fact or protocol to the project is made clear with less effort for the reader.

The well-written application thus is constructed upon an outline that serves as a skeleton for the meat of the content. This outline is indicated by the use of strong topic sentences that serve both to introduce and summarize all the information in the following paragraph. Many writers purposefully stress this format by emphasizing each topic sentence (e.g., underlined). This makes the important lines of thought immediately evident even to a quick perusal, and allows easy recall of the key points and reasoning upon later rereading. Each topic sentence therefore serves as a self-contained step on the logical path that leads the reader down your line of reasoning. Reading only the topic sentences in isolation should convey nearly the full thrust of the project (with the bodies of the paragraphs supporting and fleshing out these sentences). A key to maintaining a logical flow with the outline format is the use of smooth transitions to tie your ideas together. When several sections of text relate to a central theme, transitions between the sections should tie them to each other and demonstrate a methodical chain of thought. This gives the application cohesiveness and prevents it from seeming disjointed, with the subunits meaningfully related to each other. Here is an example of a few paragraphs written without a clear outline structure:

The earliest writing on paper is attributed to the Chinese, who invented paper and used animal hair brushes. In the more recent past, bird feathers cut to a point were used (the quill), and in the 1700s, these were supplanted by steel nibs in wooden or metal holders, to be dipped in ink. Pen lore suggests that an insurance salesman named Lewis Edson Waterman was closing the deal on an important insurance sale, when his pen leaked ink and ruined the contract, costing him the sale. Whether or not this tale is true, Mr. Waterman produced the first widely successful modern fountain pen design and started the Waterman Pen Company in 1884.

Early "dip pens" required frequent dipping into ink, because the amount of ink retained on the nib was only enough for a word or two of writing. The nibs were longitudinally slit at the point, to allow capillary flow of ink to the paper. This inconvenience led to the search for a pen design that would hold a self-contained supply of ink. Three major barriers prevented this goal. First was the fact that early inks, such as India ink, were particulate, and prone to clog any capillary flow mechanism. This problem was solved by the mid-1800s. Second was the susceptibility of steel nibs to deterioration, especially because inks were corrosive. This problem was solved with the advent of gold alloy nibs, also in the mid-1800s. Third, and most significant, was the physical problem of ink flow. Ink held in the hollow barrel of a pen would not flow properly at the slow and controlled rate required for writing, or would gush out unexpectedly (much like fluid in a capped straw) through the capillary space between the "feed" and the nib. Waterman's patented design involved cutting thin channels ("fissures") in the feed to allow minute quantities of air to replace the ink used while writing, while being small enough to maintain the closed system to hold ink. This nib and feed design has remained essentially unchanged since its inception.

The concept of a "self-filling pen" soon became popular, starting with a patented design by the Conklin pen company around 1905. Self-filling pens held ink in a compressible rubber bladder, and filled by sucking ink up through the nib and adjacent capillary channel. The earliest pens, including those of Waterman and his rival George S. Parker (who started the Parker Pen Company in 1891) were hollow tubes that needed to be opened (by means of a threaded seam in the barrel) to fill with ink using an eyedropper. This was a messy procedure that almost always ended with inky fingers. Different companies had various designs (to circumvent each others' patents), but all designs shared in common a mechanism to compress the bladder, expelling the air, leading to filling with ink when the bladder spontaneously re-expands.

Here is the same information organized into a strict outline format:

Predecessors of the modern fountain pen. The earliest writing on paper is attributed to the Chinese, who invented paper and used animal hair brushes. In the more recent past, bird feathers cut to a point were used (the quill), and in the 1700s, these were supplanted by steel nibs in wooden or metal holders, to be dipped in ink. The nibs were longitudinally slit at the point, to allow capillary flow of ink to the paper. Early "dip pens" required frequent dipping into ink, because the amount of ink retained on the nib was only enough for a word or two of writing. This inconvenience led to the search for a pen design that would hold a self-contained supply of ink.

Technical barriers to the modern fountain pen. Three major barriers prevented this goal. First was the fact that early inks, such as India ink, were particulate, and prone to clog any capillary flow mechanism. This problem was solved by the mid-1800s. Second was the susceptibility of steel nibs to deterioration, especially because inks were corrosive. This problem was solved with the advent of gold alloy nibs, also in the mid-1800s. Third, and most significant, was the physical problem of ink flow. Ink held in the hollow barrel of a pen would not flow properly at the slow and controlled rate required for writing, or would gush out unexpectedly (much like fluid in a capped straw) through the capillary space between the "feed" and the nib.

Invention and marketing of the first modern fountain pens. Pen lore suggests that an insurance salesman named Lewis Edson Waterman was closing the deal on an important insurance sale, when his pen leaked ink and ruined the contract, costing him the sale. Whether or not this tale is true, Mr. Waterman produced the first widely successful modern fountain pen design and started the Waterman Pen Company in 1884. Waterman's patented design involved cutting thin channels ("fissures") in the feed to allow minute quantities of air to replace the ink used while writing, while being small enough to maintain the closed system to hold ink. This nib and feed design has remained essentially unchanged since its inception.

Early development of ink filling mechanisms. The earliest pens, including those of Waterman and his rival George S. Parker (who started the Parker Pen Company in 1891) were hollow tubes that needed to be opened (by means of a threaded seam in the barrel) to fill with ink using an eyedropper. This was a messy procedure that almost always ended with inky fingers. The concept of a "self-filling pen" soon became popular, starting with a patented design by the

Conklin pen company around 1905. Self-filling pens held ink in a compressible rubber bladder, and filled by sucking ink up through the nib and adjacent capillary channel. Different companies had various designs (to circumvent each others' patents), but all designs shared in common a mechanism to compress the bladder, expelling the air, leading to filling with ink when the bladder spontaneously re-expands.

Attention to details

There are several further details that demand the attention of the writer, that are scrutinized by the reviewer. Beyond organization, the presentation is affected by the qualities of writing and graphics. Words are the vehicle for your ideas and plans, and the way you use them (including spelling, grammar, and syntax) determine the clarity, precision, and efficiency with which your research is conveyed. Figures are the most direct method of demonstrating your work. Legibility of the symbols, lines, and legends is essential to the point being made. Keep in mind that all figures and tables should be clearly reproducible by electronic scanning or photocopying, because NIH makes electronic copies for all members of the study section that will score your application. In the case that you must use photos or color images, you should include original copies in the appendix, which is provided to the main reviewers.

Layout

Finally, and least tangibly, the layout of the text and figures can affect the presentation and impact on the reader. Having paragraphs and graphics neatly organized with clean and even margins gives the application a conscientious and detail-oriented feel, and can complement and augment the organizational structure of the outline. Placing figures near the referring text prevents the inconvenience of flipping pages back and forth while reading.

Getting a fresh perspective

As the writer and researcher, you will be so immersed that it may be difficult to keep a fresh perspective and gauge how your writing affects other people. This makes calibrating your writing style to outside readers very difficult. An extremely helpful solution to this problem is to give working drafts of your application to colleagues who work outside your specific area of research. Their comments can be invaluable in identifying text that is difficult to understand, gaps in your explanations, problems with your organization, ambiguities in your reasoning, and other details (grammar, syntax, spelling). Areas of your application that are difficult for them to understand are likely to be trouble spots for reviewers too.

Common errors

Using small font or font compression to save space

Even if smaller font or compressed font passes administrative scrutiny at CSR, reviewers find this text difficult to read in comparison to other grants that are clearer. This is a common annoyance for reviewers, and tends to set a negative impression up front. If your application is running so long that it exceeds space limitations, the project may be too broad or inefficiently explained.

Having long paragraphs that contain too many ideas

Self-contained paragraphs that cover one or a few closely related ideas have the most impact. Long paragraphs with running commentaries may seem smoother to you as a writer, but they require far more analysis and effort on the part of the reader. If each paragraph begins with a highlighted topic sentence that summarizes the rest of the paragraph, it becomes very simple for the reader to retrieve the relevant information.

Unnecessary content

Even if a digression makes an interesting point, it may actually detract from the overall project if it obscures a major line of reasoning in the justification or plans of the project. Also keep in mind that anything you write is a potential point of criticism, so you should avoid statements that don't directly contribute to supporting your application.

Illegible figures or tables

Due to the strict guidelines on font size, margins, and page number, some writers minimize the size of their figures and tables to save space. This often results in loss of clarity, particularly when the symbols or text are reduced to illegibly small size. Large, clear figures or tables are pleasing to the eye and make the application feel less crammed.

Poorly designed figures or tables

Especially when a figure is complicated, this can make interpretation of the data a major chore. The best figures and tables convey their point when viewed at a glance; if a reader must look hard to absorb the message, it is not clear enough. There are two major common deficiencies in the design of figures. One is that too

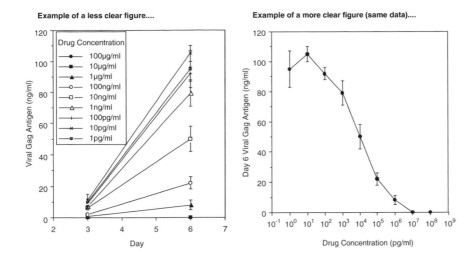

much information is presented, and that there are so many data points or comparisons that the main point is buried. Data that are clear to you, being intimately familiar with the topic and experiment, may be confusing to a reader who is seeing it for the first time. Second is poor explanation in the legend. The legend should convey the methods and experimental comparisons in a self-contained manner. While the full details may be absent, a reader should be able to examine the figure in isolation and have a clear idea of why the experiment was performed, how it was performed, and what it showed. As a general rule, a well-designed figure and its title convincingly conveys its major point to the reader within seconds.

Jargon

You should refrain from using technical jargon that is not standard accepted language. Avoid jargon or abbreviations for which the reader will need to scour the text to find. It is not worth saving a few characters of typing if the reader becomes frustrated with an unfamiliar term. Be careful to define all abbreviations or novel terms clearly and prominently. If it is necessary to use an abbreviation for clarity, you should redefine it in each section, so the definition is easily found.

Errors in the text

Numerous errors can lend an air of carelessness or incompetence. Spelling and typographical mistakes indicate sloppiness in the preparation application, and it is common for reviewers to wonder why an applicant would be so negligent as not to run the text through the spell-check function on their word processor. Poor grammar and syntax turn off many reviewers, because verbal ability is associated with other forms of competence. Thus, it is prudent to have your text carefully edited, particularly if English is not your primary language.

Crowded text layout

Even if the font and figures are appropriately sized, crowding can impair the legibility of the application. The appropriate use of blank lines and indents to separate paragraphs, subsections, and sections is important to keep them distinct to the eyes and mind of the reader. Distinctness enhances the overall organization and clarifies the structure, allowing the reader to find individual sections more efficiently. Leaving out blank lines and indents to save space often comes with the price of blurring these important aspects. Crowding also affects the visual appeal of the text, and can make the application seem unnecessarily hectic and dense.

7

Specific Aims

Introducing your project

The scientific portion opens with this section. Usually it is about one page, and contains no citations because the information is general in nature. In essence, this should be a highly condensed version of the next section, the "Background and Significance" section, ending with a listing of the actual "aims" of the application. As a rule, it is a synopsis of the information that explains to the reader why your proposed research will be important to pursue, and the context into which the work would fall. Once the reader has read the remainder of the grant application, he or she should be able to reread this section as an effective memory refresher for why the work is important and what questions it will answer.

Summarizing the rationale

The "Specific Aims" section begins with background information that presents the rationale for the application. This information should provide a broad context appropriate for a reviewer who may be a researcher in a related but separate field. The pertinent facts about what is known and unknown in the area of the research, the relevant questions to be addressed about this area, and the potential implications and ramifications of the work should be summarized in two or three short paragraphs. These paragraphs should be self-contained and should stand alone. Obviously, such a short synopsis will not explore all the fine nuances in the research area of interest, but it should provide a coherent explanation to a scientist working in a related area. The following questions should be addressed:

1. What is the field of investigation?
2. Why is it important?
3. What is the state of knowledge in this area?
4. What important questions remain unanswered?
5. What would be the ramifications of answering these questions?
6. What limitations have posed barriers to research in this area?
7. What new or potential developments offer you a successful avenue to address your questions?

The end of the final paragraph should consist of a lead-in sentence such as "To approach these goals, we specifically propose:" followed by a listing of the actual "aims" for the project.

Stating the aims of the project

There should be two to four aims that are succinct statements precisely summarizing the research to be performed. Each aim should cover one basic research focus central to the topic of the grant. They should be thematically related and fit together to clarify the important issues raised in the introductory paragraphs. They may each explore a different idea, or different approach to the same idea, as long as they form a cohesive thematic unit, addressing an overall central idea or hypothesis. Each aim should be headed with a number ("Aim 1," etc.), to which later parts of the grant will refer.

Types of aims

The aims are one of two types. The first is the "hypothesis-driven" aim, based on a scientific hypothesis to be tested and substantiated or refuted by the data that will be generated in your proposal. The hypothesis should be reasonably justified by observations, either published or personal (given in the "Background and Significance" and/or "Preliminary Results" sections) that have interesting implications. This hypothesis-driven approach is favored, because it is immediately apparent what would be gained from the proposed work. The second type is the "exploratory" aim, which does not carry a specific hypothesis, but seeks to obtain useful data about a specific question. This type is descriptive and emphasizes gathering information or taking a "fishing" tact to allow future formation of hypotheses.

For larger grant applications by independent investigators, such as R01s, the proposal should contain at least one major hypothesis-driven aim, and preferably be mostly or all hypothesis-driven. Such grants are intended to fund a focused and targeted attack in a research area, and thus should contain distinct testable hypotheses. However, it may be entirely appropriate for a portion of such a project to be exploratory as well, as long as the grant is not primarily exploratory. For smaller type grants such as R03 and R21 applications, a predominantly exploratory focus may be appropriate, because these projects are intended to be preliminary and higher risk. Still, it is useful to incorporate hypothesis-driven work when possible even in these projects, as this gives the proposed work direction and focus. Even if an aim is exploratory and not hypothesis-driven, it is crucial to justify the rationale for how the information acquired would open an important field of investigation and lead to new hypothesis-driven work.

Giving further details on the aims (or not)

As a matter of preference, some authors give additional brief descriptive information under each aim. In this case, a brief paragraph is provided under each listed aim, to summarize the hypothesis and approach to be taken. If this is done, the information should be sufficient to stand alone, and should not be cluttered with details (particularly methodologic) that distract the reviewer from the central point of the aim. Because the "Specific Aims" section should be very concise, it may be better to restate the aims with these additional details at the end of the "Background and Significance" section (see Chapter 8).

An example

For example, the "Specific Aims" section for a hypothetical grant addressing "Idiopathic acute pan-cerebral necrosis" (not a real disease) could begin with a paragraph explaining the incidence, prevalence, and impact of this disease, continue with a paragraph concerning the epidemiologic data, and finish with a paragraph on the specific scientific facts that are pertinent to the aims, such as description of the available models and current thinking on pathogenesis of this disease, concluding with:

> ... In light of our recently enhanced understanding of the neuron explosion cascade (NEC) and its likely role in Idiopathic Acute Pan-Cerebral Necrosis (IAPCN), identifying abnormally expressed neuronal factors is likely to shed light on the pathogenesis of IAPCN. The central focus of this application will be to apply newly developed ultrasensitive techniques to identify abnormal genetic and protein expression of neuronal factors and investigate the role of these factors in the development of IAPCN. Specifically, we propose:
>
> 1. To quantify and localize expression of NEC factors in an established murine model of IAPCN;

2. To evaluate the in vivo effect of small molecule inhibitors and enhancers of NEC components in the murine model of IAPCN; and
3. To assess the metabolic activity in relevant brain areas in humans at risk for IAPCN.

As the opener to the science portion, this section will set the first impression for the reader. The reviewers will read it carefully, because it serves as an orientation for the rest of the grant application. This impression thus will be especially important to point the reader in the right direction and set the tone for the rest of your application.

Common errors

Unrealistic aims

Because success in meeting the aims is the ultimate goal, they should be realistic. They should be "specific" and focused aims, and not so broad that they make unsustainable claims such as understanding the entire pathogenesis of a disease. The reviewers should see how the aims could be accomplished in the lifespan of the grant. Additionally, the aims usually should be feasible given the technical expertise displayed in prior publications or in the application. However, in certain cases feasibility may be less important. In larger grants, subsections or an entire aim may be high risk, as long as the potential payoff of success is great and failure does not detract from the overall value of the project. Another case where feasibility may be less important is in grants that are intended to be exploratory (e.g., R03 or R21) or for training purposes (e.g., K08), in which cases the overriding criteria are the potential benefits and training provided by the project, respectively. Even in these latter cases, it is important to justify why the risk is worthwhile, and how the project could be useful even if the aim(s) are not entirely met.

Poorly justified aims

If it is not clear to the reader why the current state of the field justifies the aims in terms of importance, relevance, or support for hypotheses, the rest of the application is thrown into question. Particularly in the case of grants responding to a Request for Applications (RFA) or Program Announcement (PA), justifying the aims in the context of the RFA or PA goals is crucial. The relationship of the aims to what is known, and what needs to be known, should be made obvious to the reader, and the hypotheses should have clear support.

Purely descriptive aims

Although portions of a larger grant (e.g., R01), higher risk grants (e.g., R03 or R21), and training grants (e.g., KO8) may be exploratory in nature, a purely descriptive tone should be avoided. Common criticisms are that the described work has a "laundry list" approach, or that it appears to be a "fishing expedition." When possible, it is advantageous to have aims that are hypothesis-driven. If the work is necessarily exploratory, the context for why an empiric approach is required should be made clear, and the potential ramifications (including possible resulting hypotheses) should be explored where possible. You want to avoid the impression that the project is simply a list of experiments because you cannot devise a targeted approach.

Unnecessarily complicated aims

Each aim should be focused on a clear idea and serve as a thematic unit. Some writers divide aims into multiple "subaims" listed directly under the general aim. This is a matter of preference and personal style, but if an aim requires division into several subaims, it may lack the proper focus. Having single aims is probably the more effective method to keep the individual ideas clear in the mind of the reviewer. The "Specific Aims" section should paint the picture in broad strokes, with the finer details to come later.

Lack of cohesiveness of the aims as a unit

The aims should be thematically related, and form a cohesive unit when possible. Most reviewers look for a "central hypothesis" that unifies the aims. Jumping from one idea to an unrelated idea gives the application a haphazard feel. Although the aims themselves should obviously address distinct ideas or principles, their relationship to each other should be clear, and the research plan as an entire unit should be cohesive. Ideally, the individual aims should dovetail and be synergistic.

Excessive interdependence of aims for success

A common error is having aims that are entirely dependent on the success of earlier aims. If an aim depends on reagents or protocols that don't yet exist, its success becomes particularly dubious. The technical success of any aim should generally avoid requiring the success of an earlier aim to generate reagents or protocols, because this severely risks success of the project overall if just one aim fails. For example, an aim to study the effects of monoclonal antibodies in an animal disease model should not depend on an earlier aim to generate those antibodies, unless that process is so absolutely straightforward that the reviewers would not doubt the ability to do so.

8

Background and Significance

Laying the foundation for your project

This section serves as a foundation for your scientific ideas and research plans. It is usually two to four pages, and contains important facts backed by literature citations. These facts should lead the reader down the logical path to concluding that the specific aims are worthwhile to pursue. For grant applications that are written in response to an RFA or PA, it is particularly important that this section address the relevance of your project to the stated goals of the RFA or PA. The "Background and Significance" section, when well written, should serve as a stand-alone review article that gives the expert reader all the necessary information to explain:

Why the area of research is important. The rationale for your research should be clearly delineated. The scientific issues you address should be tied to some important fundamental problem, for example, pathogenesis or prevention of a threatening disease, a poorly understood basic molecular process used by many bacteria or viruses, a potential explanation for a phenomenon of unknown mechanism. The reviewer should be convinced that the aims, if achieved, would contribute to a fundamentally important topic with implications that unravel a key unanswered scientific question, or potentially provide a new clinical avenue. To continue the fictional example in the previous chapter, delineating the pathogenesis of IAPCN may be important because this rare disease shares important pathologic principles with Alzheimer's or Parkinson's Diseases and light will be shed on these more common syndromes. Perhaps IAPCN is a dysregulation of normal neuronal death during brain development, and understanding the processes will shed light on poorly understood central questions about brain ontogeny. Perhaps IAPCN is a localized form of

Spontaneous Human Combustion Syndrome, which has been a central mystery to combustionologists all over the world. Perhaps lives can be saved because, although rare in the general population, the incidence of IAPCN is high among new assistant professors, and the research will lead to useful preventive measures or therapeutics that will have far-reaching effects on society by saving this important group of people. Although scientific knowledge is important and interesting for its own sake, your research will carry more weight if you can tie it into some larger scientific or clinical picture (even if the work does not have immediate application).

What is known thus far about the area of proposed research. Work pertaining to the area of interest should be summarized and clearly cited in a comprehensive manner. This includes scientific discoveries leading up to the questions being asked, previous attempts to address similar questions, current theories, previous limitations that have prevented resolution of the area, and recent advances that may assist in pursuing the aims. There should be sufficient information here to educate the reader about how your work would fit into the field and advance it. The information should be comprehensive and clearly organized, with citations crediting anyone who has contributed to the field (see also Chapter 11). This review should be as objective as possible, and avoid subjective statements that might alienate a reviewer with an opposing view. Instead, it is better to give the clearly established facts, acknowledge controversy, state the prevailing hypotheses in the field, and add your interpretation of the facts with reference to supporting data. Focusing on the positive (why your viewpoint could be correct) is always better than the negative (why others are incorrect). The goal is to convince the reviewer that you are thinking clearly and objectively about the issues, and have a firm grasp of the field.

Implications of the proposed work in the field. A brief explanation should be given as to how accomplishment of the aims would broaden or change the field. This should be related to what is already known, and how your data would enhance or change the current state of knowledge. If applicable, mentioning potentially useful applications of your findings is also helpful, although great care should be taken to avoid making claims that are tenuous or unsupportable.

Organization and Structure

As discussed earlier, an effective writing style is to break this section into multiple paragraphs that follow an outline form, with the topic sentences underlined. This is particularly important for the "Background and Significance" section, where the reader needs to assimilate a large amount of information quickly. Each topic sentence should convey the entire point of its paragraph, with the other sentences elaborating on the data or rationale for that statement, making the key points easy to absorb at a glance. This provides the reviewers with visual cues for quick recall of the information (that they may have read weeks earlier) when skimming the application during the review meeting.

An example

Some authors conclude this section by listing the aims again, in the context of what has just been described. The restatement and expansion of the aims here can strengthen this section by directly tying your proposed research to the information presented. After each aim, a few sentences now can be given to summarize the experimental approach and how it relates to the rest of the background information. To continue the fictional example from the last chapter, we could conclude the "Background and Significance" section with:

> … Specifically, we propose:
>
> 1. To quantify and localize expression of NEC factors in an established murine model of IAPCN.
>   While the importance of the NEC in IAPCN has been clear, standard immunohistochemical methods have lacked the sensitivity and quantitative precision to define NEC dysregulation in IAPCN. Recently, our group has developed ultrasensitive methods to perform quantitative in situ localization of NEC factor transcripts and proteins. We will apply these methods to a well-established murine model of IAPCN to evaluate the regulation of NEC factors in different anatomic locations during various stages of disease development.
>
> 2. To evaluate the in vivo effect of small molecule inhibitors and enhancers of NEC components in the murine model of IAPCN.
>   Novel small molecule inhibitors and enhancers of factors in the NEC are available in our laboratory, and we have utilized them successfully as tools to study murine Parkinson's Disease. To evaluate whether downregulation or upregulation of NEC components have a causal role in IAPCN, we will treat control and IAPCN mice with these small molecules. The results will be correlated to those of the first aim.
>
> 3. To assess the metabolic activity in relevant brain areas in humans at risk for IAPCN.
>   While evaluation in the murine model will allow precise localization of NEC dysregulation, such studies are not feasible in humans. We have established a collaboration with a neuroradiology laboratory that specializes in positron emission tomography (PET) scanning to localize metabolic activity in neuropathogenesis. Our cohort of IAPCN subjects will be evaluated with PET to explore metabolic abnormalities in this disease, and the findings will be correlated to those of the first and second aims.

Common errors

Too little detail

This section should be written as a self-contained review article that provides enough information to make your case for the research proposal. You should calibrate the writing to the needs of the reviewers. For example, if you are writing a grant application concerning the effects of ethanol on T cell signaling pathways as a mechanism of alcoholism-induced immunodeficiency, and you decide to submit to a study section that reviews grant applications concerning substance abuse, it is likely that your reviewers will not be T cell immunologists. You should therefore explain T cell signaling in detail, using minimal jargon, and carefully tie the relevance of the basic science to the clinical implications, to make your application readily accessible to substance abuse experts with little background in immunology. Another frequent failure is to abbreviate the "Background and Significance" section, assuming that the information in attached Appendices will suffice (see Chapter 12). This section should be self-contained and tell an interesting story, without expecting the reader to search for information elsewhere.

Too much detail

At the other extreme, it is possible to deluge a reader with too much detail. If the same grant application above were going to a study section specializing in immunology, giving laborious detail about the definition of the T lymphocyte and why T cell signaling is important might be too tedious and lose the interest of the reviewers, who will have expertise in T cell immunology. As mentioned above, calibrating this section to the readers is crucial.

Poor organization

Even if all the relevant facts are included in the "Background and Significance" section, they will not make sense unless properly organized. The outline format is the best organization to make all the important information rapidly accessible, because the facts are clearly filed under topic sentences that support your thinking. Filing all supporting evidence for each idea under a clear topic sentence makes it simple for the readers to synthesize the data and reach the same conclusion as you.

Lack of objectivity

Although the purpose of the "Background and Significance" section is to introduce and support your hypotheses, you must be careful to do so in an even-handed and objective manner. There is a fine line between being focused and being close-minded. If you cite evidence for your viewpoint but do not cite existing contradictory evidence, the reviewers will assume either that your knowledge is lacking, or that you intentionally ignore opposing evidence. Neither makes a good impression. It is always important to acknowledge competing views and interpretations, even if you disagree. It is possible (and ideal) to take a firm intellectual stand without excluding other potential theories.

9

Preliminary Results

Demonstrating the feasibility of your project

Most grant applications require this section, to present data showing the feasibility of your project. Major exceptions include exploratory grants such as the R03 and R21, which do not require showing any data. However, because it is so difficult to make a case without showing data that an interesting idea can be successfully explored, almost all applicants for these grants do include the "Preliminary Results" section. For common independent investigator grants, such as the R01, this section carries much weight in the scoring of the application, because these grants place heavy emphasis on the likelihood of successful research, which in turn requires demonstrating the ability to perform the work proposed. For training grants such as the K series, data in this section may come from the mentor's laboratory, since it is not necessarily expected that the trainee would have done much research. When data have been generated by the trainee, it is greatly advantageous to state this explicitly.

Purposes

The "Preliminary Results" section has two major purposes: to support the scientific basis for the aims of the grant, and to demonstrate that you possess the technical capability to carry them out. It is useful to organize it much like the results section of a paper, with focused observations in individual subsections. To make it easy to skim and review (like the "Background and Significance" section), each subsection should be headed with a sentence or phrase that concisely conveys its observation and conclusion. These headings should be emphasized with underlining or bold font, to stand out visually and facilitate rapid reading. Again, the more easily absorbed and recalled the information, the clearer and more durable the impression is on the reader.

Amount of data to include

In general, larger projects (such as the R01) require more detailed data that are directly relevant to the goals of the proposed project, and smaller projects (such as the R03, R21, and K awards) are less stringent and more focused on technical feasibility. Again, calibrating your writing to the goals of NIH is important. The emphasis of larger grants is producing meaningful results, and these therefore require more data to support the concepts and approach in the proposal. The emphases of smaller grants are high-risk exploration or training of new scientists, and these therefore require less support for the concepts and more support for the techniques to be applied or learned. Even in the latter, however, providing data that are relevant to both concepts and techniques is preferable.

Format

Each subsection should contain data in the form of figure(s) or table(s). The data should either support a hypothesis of the grant application, show a technique to be used in the proposed experiments, or ideally both. It is very important that the data should be relatively self-explanatory, and visually oriented to convey the point quickly and clearly. A brief legend should be sufficient to explain the idea of the experiment, without being cluttered by too many details. The more easily and quickly absorbed, the greater the impact the information will carry with the reviewers. The quality of the data should convince the reviewer of your ability to perform the experiment without needing to explain the minutiae of the techniques, particularly for standard accepted methods. The bottom line is showing that the tools to approach the aims are in place and functional in your laboratory or through close collaborators.

Supporting hypotheses and demonstrating techniques

For developed projects such as R01 applications, the key hypotheses should be addressed by data. Evidence should be presented that shows your hypotheses are plausible, feasible, and experimentally approachable. The sum of the data should tell a cohesive story that steers the reader toward understanding the reasons behind the hypotheses, and hopefully whets an appetite to find out more. Reading the "Preliminary Results" section should escort the reviewers down the logical path of your research interest, and point to the next steps that should be taken through your aims. This path should be clear and easy to follow, for a reader who is knowledgeable in the general research area, but not necessarily an expert on the specific topic.

Methods central to the grant proposal should be demonstrated if possible. Experimental techniques should be described in enough detail to convince the reviewers that you can carry them out, but not to the point that excessive time and space are spent describing obvious procedures. The more novel or unusual the technique, the more detail will be required. For example, standard DNA hybridization gel blots require very little explanation, while a new nonstandard technique such as protein-to-RNA reverse translation would require precise explanation to prove feasibility. The more unusual or novel the technique, the more important it is to demonstrate and give details in this section. Finally, sources of unusual reagents should be mentioned, to document your ability to procure them for the work. The level of detail should also be appropriate for your level of training; if you are writing for a training grant application, more detail about standard techniques is required than if you are a seasoned investigator writing an R01. In short, the reviewers need to be convinced that you possess or can definitely acquire the abilities, techniques, equipment, and reagents to perform all the experiments listed in your experimental plans in the next section, "Research Design and Methods."

Organization

It is useful to conclude each section with a sentence that summarizes the point of the section. While it may seem redundant, this makes it obvious to the reviewers how the information is pertinent to the project as a whole. This directness reinforces and unifies your ideas, giving the whole application cohesiveness. Keep in mind that the reviewers will be reading multiple applications, and saving them time by plainly stating the relevance of your points will make reading your application easier. It will also facilitate rapid retrieval of information if the primary and secondary reviewers need to refresh their memories later, or if another reviewer wants to find a fact without having read the application in full detail.

To continue our fictional example of a grant application concerning IAPCN from the previous chapters, a section in the "Preliminary Results" could begin:

> In situ quantitative PCR of neuron explosion cascade factors.
>
> We have applied a novel in situ quantitative PCR method to study the role of NEC factors in a murine model of Fatal Familial Anhedonia (Yang and Pavlov, 1999) ....

After presenting relevant data demonstrating this novel technique, the section could then conclude:

> ... Because it recently has become clear that the NEC is an important mediator of IAPCN, we intend to apply in situ quantitative PCR to study the

localization and expression levels of NEC factors in an established murine model, using a modification of the above approach we applied to studying FFA.

Common errors

Failure to indicate the source of data

All data should be provided directly from your work or collaborators on this specific application, because this section is intended to show the capabilities directly available for the project. Other data relevant to supporting the hypotheses of the application belong in the "Background and Significance" section, and not this section. The source of the data should be clearly indicated, to identify whether it comes from your laboratory or your collaborators. For training grants (F and K series), it is expected that much or all of the data come from the laboratory of the mentor, although data provided by the applicant is especially valuable. For grants to established investigators, the bulk of the data should come from the applicant's own laboratory.

Unnecessarily complicated figures or tables

Clarity is the key issue. A figure that contains multiple arms making multiple points is difficult to absorb, and can be frustrating to the reader. Each figure should convey one key idea and focus; a complicated figure is better if broken into multiple parts or separate figures. If numerous arms are needed as controls, it may be useful to show them in a separate figure, or to refer to them as data not shown (especially if you can cite a publication or appendix that shows those controls).

Showing data not clearly related to the proposed work

All information presented in this section should contribute to understanding and substantiating your hypotheses, and/or showing feasibility of the experiments you propose. Information that is peripheral (however interesting) detracts from these goals and expends time and effort of the reader needlessly. Also remember that the reviewers will be reading this section before seeing your experimental plans, so it is important that you emphasize the relevance of each section in a manner that ties it to the project, and sets up the reader for the plans ahead. This section should provide a crucial transition between the "Background and Significance" and the "Research Design and Methods" sections. If the content is not relevant but the methodology is pertinent, briefly explain how the methodology will be applied to the research questions at hand.

Poor organization

When well written, the "Preliminary Data" section has a logical flow; otherwise it can seem like a random assortment of experiments. The better the components fit together as a unit, the easier it is for the reader to put all the information into a meaningful context that will apply to your research plans. Organizational skill demonstrated in this section provides a positive reflection of your ability to think and assemble data. Having logical transitions between subsections and putting them into an order that tells a story can markedly enhance the digestibility of the information, and make it more interesting.

Omission of important points or procedures

New ideas or techniques that will be utilized in the experimental plans should be addressed in the "Preliminary Data." Supporting data should be presented here; otherwise, a novel hypothesis or method in the "Experimental Design" may be criticized for lack of feasibility. If the idea or technique is central to an aim, showing data is usually essential to convince the reviewers. Simply referring to a publication, even an appended paper, is inconvenient to the reader. This section should stand alone, without requiring any supplemental information (although you may refer the reviewers to appendices for specific details). The desired final result is convincing the reader that all the tools and ideas are in place for your project, without packing the section with minutiae.

10

Research Design and Methods

Describing your experimental plans

This is obviously the heart of the application and will receive the most scrutiny during review. The emphasis of the reviewers will vary somewhat depending on the type of grant. For training grants (F and K series), this section will be judged primarily on whether these experiments will be a useful exercise in learning skills through performing research; for independent investigator grants (R series), the emphasis will be on the appropriateness of the plans to achieve the stated specific aims. In both cases, this section should be highly organized and clear.

Importance of logical organization

The "Background and Significance" section describes the research plans for the entire project in a format that clearly conveys to the reviewers your plan of action to tackle the specific aims you have proposed. Experiments to address your hypotheses should be organized in a logical manner that displays a methodical approach to the scientific issues, with sufficient detail to convince readers of feasibility. This section should be a roadmap for research during the years of the proposed project, and clearly show the reviewers how you intend to proceed, with alternative routes if you hit roadblocks.

It is advisable to begin with an "Overview" subsection of a paragraph or two. This should serve as orientation to the research as a whole, and present the big picture. This overview should integrate the questions posed by the all the specific aims, after summarizing the "Background and Significance" section in a few sentences to serve as a reminder and transition for the reader. The overall approach for each aim should then be painted in broad strokes. Reading this subsection should provide a "forest" context for the "trees" to follow.

Each specific aim is then covered in an individual section, beginning with a verbatim restatement of the aim. The information provided in this section should elucidate the idea being explored, the specific experimental approaches and protocols, and the plans for interpreting the results. Different authors follow various organizational schemes, but including the following subheadings under each aim ensures coverage of all the key elements.

Subsections under each aim

Hypothesis. This is a succinct (one or two sentence) statement of the hypothesis being explored in the aim. This statement should be a clear expression of a theory, and not a question or a vague description of an area to be explored. You should articulate the hypothesis alone without supporting statements here; the evidence should be clear from the "Background and Significance" and "Preliminary Results," sections as well as being briefly summarized in the following "Rationale." Lengthiness defeats the purpose of providing an easily remembered central idea here.

Rationale. This is a concise (one or two paragraph) explanation supporting the hypothesis. High points among the evidence for the hypothesis (already covered in the "Background and Significance" section and Preliminary Data) should be summarized, so that reviewers can appreciate the thought process without being bogged down in details (that are available in the other sections). The information should be succinct and efficiently convey a reasonable case for the hypothesis.

Experimental Approach. This section describes the experiments to be performed to address the aim and its hypothesis, and is the meat of each described aim. It should be further broken into subsections that contain separate experiments or procedures. Important reagents and/or their derivation may warrant their own subsections, if they pose a potentially significant hurdle to the research. The degree of detail to include in each subsection should be as brief and clear as possible to avoid distracting the reader with trivialities, but maintain enough information to convince the reader that the approach can work and that you have the requisite technical ability and reagents. Standard assays in common usage may be abbreviated, especially if they are demonstrated successfully in the "Preliminary Results" section or an appendix, but any experimental procedures that are nonstandard should be described in detail. Methods that are not already successful in the laboratory should be supported with enough information to convince the reader that they can be achieved readily; either demonstration of closely related techniques in the "Preliminary Results" section or documentation of technical support from a collaborator are usually needed. The sources of all key reagents should be made clear, for example, by listing catalogue numbers/company names, or by referring to appended letters of support from collaborators. The listed experiments should fall into ordered categories and have a sensible flow, leading the reader through an organized train of thought to approach the issues raised in the aim. For complicated experiments that involve multiple arms or repetitions under different conditions, it is often clearer to show a table that demonstrates the various permutations, rather than trying to explain them through a running commentary.

Interpretation of Results. This is a crucial section that is often either entirely overlooked or underdeveloped by many grant writers. Although it may be obvious to you what the experimental results would indicate, it is often not so clear to the reviewers. Whereas you (as a researcher in this area) have these issues in mind constantly, it is important to remember that the reviewers may be thinking about this topic for the first time, and the implications of the experiments you propose may not be intuitively clear to them. Thus, even the most elegant experimental plans can have little impact unless their interpretation is clearly delineated. In fact, especially novel approaches may be paradoxically perceived as irrelevant if their link to the hypotheses is not made obvious.

Depending on the results expected, this subsection may be very brief or require detailed explanation. Generally the more tightly focused and obvious the aim, the less explanation will be required. For example, if you have already explained in the "Hypothesis" and "Rationale" subsections why it is important to identify binding partners for a particular protein, and your aim is to identify these partners through several techniques, a brief comment about anticipated results may suffice. However, if your aim is to perform a multivariate analysis of a patient population, it may require many paragraphs to explain the meaning of the many possible findings. The typical "Interpretation of Results" subsection falls between these extremes, and runs one to three paragraphs. Finally, a crucial yet often neglected consideration for this subsection is the possibility that a hypothesis presented for this aim is incorrect. Provision should be made for this possibility, and it should be made clear that the interpretation of the data will be able to distinguish this circumstance and allow revision of the hypothesis.

Potential Pitfalls and Alternative Approaches. This subsection is often omitted altogether in many grant applications, despite being perhaps one of the most important in the eyes of reviewers. It is almost never appropriate to leave this out. Here you discuss possible roadblocks to your experimental plans, and how you would adjust your approach to get around them. Being realistic about your limitations is crucial here, and you should show awareness of the potential problems you face in your plans, with an appropriate balance of confidence and caution. Being able to anticipate the criticisms of a reviewer and to trump them with foresight and discussion of alternatives is an opportunity to demonstrate how clearly you have thought through your project. Failure to see obvious weak points in your experimental plans can give the impression that your approach is careless and therefore less likely to be successful.

Even if you have considered all the pitfalls during the design of your experimental plan, this will not be apparent to the reviewers unless you explicitly discuss them. Regardless of whether the plans for the aim are obviously achievable or fraught with risk, this subsection should be included. Even if the experimental approach for the aim is entirely standard for your laboratory and all the reagents are already in your hands, it is appropriate to make a short statement to this effect to indicate the importance of these issues in your mind. If the experimental plan is unavoidably high risk, discussing the possible points of failure and why this is still the best approach will show the reviewers that you are not blindly pursuing this path, and that it is a worthwhile approach to a difficult question despite the risk. This will also prevent the reviewers from assuming that you are taking this tact because you cannot see the problems it entails. The key is to reveal your thought process, and ability to troubleshoot.

One major pitfall that should be addressed, if appropriate, is the possibility that a hypothesis being explored is incorrect. As in the "Interpretation of Results" subsection, this possibility should be given clear consideration. Here it is pivotal to explain how the experimental plan would be modified or the research direction changed if this is the case; otherwise the entire aim loses its validity if the hypothesis is wrong.

Overall Summary and Significance. After the final aim, many authors end this section with a subsection entitled "Overall Summary and Significance." This is a concluding statement that revisits the big picture and discusses how the results from all the aims pertain to the general questions posed in the grant application. Depending on the complexity of the topic, this discussion may be one or several paragraphs. It should not be unnecessarily verbose, but should concisely provide a clear context for assessing the importance of the work. The importance of the results and the impact on the field should be discussed, concluding the section by stressing the relevance of the work.

Common errors

Lack of "Hypothesis" and "Rationale" sections

To the writer who is completely immersed in the work, these issues may seem too obvious to spell out, and he/she may feel redundant to state these points. However, to the reader who is not fully versed in that specific area, direct and concise statements of the hypothesis and rationale are invaluable for setting the context of all the experimental plans to follow in that aim. Some writers blend the hypothesis and rationale in an introductory paragraph for each aim, which is acceptable, but it is easier to the reviewers if they fall under their own clear headings.

Inadequate experience of the investigator in proposed techniques

Demonstrating experience in the proposed techniques is particularly important for larger grants by independent investigators (where a key consideration is the probability of success). The burden of proof is on the applicant. The application should be convincing to the reviewers, through a combination of prior credentials, publications, preliminary data, and collaborations. For example, if you are a newly independent investigator with a minimal track record, you will need to provide more supporting evidence in terms of data and technical detail, and perhaps documentation of collaborative assistance from an expert.

Excessive or lacking experimental detail

There is a balance between providing enough details to convince the reviewers of your ability and flooding the description with unnecessary information making the plan laborious to read. A good general rule is to keep the level of detail directly proportional to the novelty of the methodology. Standard techniques that are widely used and straightforward should not require much explanation (e.g., cloning DNA or running ELISAs with commercially available kits). At the other extreme, an assay you have recently invented may require full details (and data given in the "Preliminary Results" section) to convince the reader of feasibility.

Missing controls

The ability to interpret results is usually highly dependent on having the appropriate controls, and heavy emphasis is placed on controls in the reviews. Although it may seem tedious or too obvious to list all the relevant controls for your experiments, the reviewers are likely to criticize omissions. This error is one of the most common in this section.

Inadequate details on clinical subjects or specimens

Enrollment of human subjects or availability of samples is commonly a limiting factor in clinical or translational studies. A frequent criticism of experimental plans utilizing clinical specimens is insufficient detail concerning the numbers that will be available. Showing the reviewers that there will be enough clinical material to produce significant findings is important for such studies. A very common error is failure to provide information concerning availability of samples. Giving exact numbers of samples in hand or justified approximations of samples that can be obtained, in concert with statistical power calculations to estimate the required number of samples to give significant results, goes a long way toward satisfying this common criticism.

Lack of "Pitfalls and Alternative Approaches"

This information is often neglected, despite being extremely important. Obviously, unforeseen obstacles can thwart your research plans, or your hypothesis may be incorrect. A major concern of the reviewers is that the entire aim or project falls apart in these cases, resulting in a waste of time and resources. Being able to propose alternative approaches to your goals if your plan fails, or new directions to move your work if your hypotheses need revision, scores a big impression in demonstrating that your research will yield meaningful results and showing your ability as a critical thinker.

Overly ambitious experimental plans

Because achieving the aims is the ultimate goal being reviewed, the feasibility of the experimental plans is a key factor. A very common error by new investigators is proposing plans that are too ambitious in scope or volume. The application should serve as a blueprint for your work, not as a wish list for experiments that seem impressive. The reviewers will consider whether the workload is realistic; if your plans are excessively optimistic, this will cast doubt on whether you can deliver results and prioritize your investigations. It may be appropriate to have high-risk goals and approaches, but the risks should be carefully acknowledged, and potential gains in the event of either success or failure emphasized. The volume of work should always be reasonable and achievable.

Plans that are not hypothesis-driven

Particularly for grants to support established researchers (such as the R01), hypothesis-driven aims and experiments are preferable. A frequent criticism for many grants is that the described work is a "fishing expedition," without clear goals or direction. Each experiment should have a clear purpose; if you can't easily articulate why it should be performed, chances are that the reviewers will find the proposal unfocused and vague. Some researchers will list every possible approach imaginable "to be complete," but this often gives the impression that the experimental plans are simply a disorganized list. It is far better to focus and prioritize crucial experiments around a clear hypothesis, and to list the rest in the "Pitfalls and Alternative Approaches."

Important exceptions to this guideline are exploratory sections of hypothesis-driven grants, or entirely exploratory grants such as R21s and others. Significant portions of these latter applications may be exploratory and descriptive. However, all proposed exploratory work should have a potential payoff that justifies the cost and risk of the exploration, and this payoff should be detailed. If you can tie the possible findings to existing hypotheses or possible new hypotheses, it greatly strengthens the case for the work.

The line between being "hypothesis driven" and "exploratory" can be very fine, and sometimes slightly adjusting the slant can put your experimental plans in a more favorable light. For example, proposing "to measure the relative levels of immunomodulatory Th1 and Th2 cytokine levels produced by infiltrating lymphocytes in IAPCN" may give a more hypothesis-oriented flavor than proposing "to utilize gene chips to measure cytokines produced by infiltrating lymphocytes in IAPCN," even though the phrases describe identical approaches. Phrasing your ideas in relation to hypotheses or concepts can be useful even for aims that are purely exploratory.

Complete dependence of the whole project on a single unproven premise

This is particularly applicable to large grants such as the R01, where the surety of success is an important aspect of its judgment. If the research plan and importance of the results all depend on a hypothesis that is not proven, this can be an "Achilles heel" of the grant. For example, an application that seeks to show that women differ from men in progression of a viral disease and then evaluate the influence of sex hormones on different arms of immunity becomes useless if the initial difference between women and men cannot be demonstrated. Well-designed research plans for big projects usually guarantee useful information, even if the hypotheses underlying the work require modification or revision. Again, a notable exception to this general guideline is high-risk innovation grants such as the R21 or R03. For these, the risk of a "hit-or-miss" approach may be acceptable if the potential payoff is high (and described clearly).

Excessive dependence of experimental plans

Aims that depend on the success of a previous aim are susceptible to criticism if the success of the previous aim is not assured. For example, if an aim gives plans to identify specific antibodies, and a subsequent aim depends on the use of these antibodies as reagents, failure of the first aim will guarantee failure of the subsequent aim. Of course interdependence of thematically related aims is usually necessary to a degree, but the experiments and aims should contain enough diversity to assure acceptable levels of success. A carefully written "Potential Pitfalls and Alternative Approaches" section can be useful to address these issues.

Lack of "Interpretation of Results"

This is perhaps one of the most common serious flaws. Some authors entirely omit an "Interpretation of Results" section. Neglecting this section is always unwise, even if the importance of the results seems intrinsically obvious to you. First, the meaning of the results may not as obvious to a reviewer. For example, determining the cytokines playing a role in a disease process may be obviously important to you as an immunologist, but obscure to a reader who is a microbiologist. Second, this section ties the experimental plans to the overall scheme of the research. Redirecting the reader to the underlying research questions and linking the experiments back to these principles gives the project cohesion and makes it a more interesting story. Third, this is an opportunity to showcase your skills in analyzing data, and putting them into context. You can demonstrate your capabilities to predict the results, consider alternative results, refine your hypotheses, adjust the experimental plan, and assess the impact of data.

Over-optimism concerning the implications of the results

The "Interpretation of Results" is an opportunity to shine intellectually, but it can also reveal shortcomings in your thinking. A common criticism is that the interpretation is overly optimistic or even unfounded. Care must be taken not to overinterpret the significance or meaning of the anticipated results, and the reasoning should be conservative. Caveats and possible alternative interpretations should be acknowledged and discussed. Presenting a balanced, open-minded perspective can prevent criticism from a reviewer who may have different viewpoints on the issues at hand.

Lack of logistical organization and justification of the work

If your project has many parts that are interrelated, or if you are requesting support for work that will take several years, it is very helpful to show the reviewers your logistical plan for completing the individual parts of the project. It is generally a good idea to provide the reviewers a timeline for completing the major portions of the aims, either as a graphical figure or table. This shows the reader your emphasis for how the work will be completed, and importantly, justifies the years of support you are requesting.

11

Use of Literature Citations

Importance of documentation

Documentation is an important aspect of the application. As mentioned earlier, the "Background and Significance" and the "Preliminary Results" sections are intended to bring the reviewer up to date in your research area and to demonstrate your technical ability to perform the proposed work. Published findings in support of both are essential, just as for a review article. The use of literature citations is therefore central to supporting these sections and putting your work into context within your field.

Format

There are no strict guidelines for the format of the citations and bibliography, as there are for specific journals. The choice of format is therefore a matter of personal style. In line with the concept that ease for the reader and clarity are paramount, many writers choose to use the "author-year" format, where in-text citations are referenced by the last name of the first author and year of the publication. For a reviewer who is somewhat familiar with the literature in your research area, this is often the easiest format to read, because this information is enough to bring the relevant work, if not the specific paper, to mind. Because the "Literature Cited" section does not apply to the page limit of the scientific section, the references (authors, title, year, journal, volume, pages) should be completely provided in the bibliography.

Demonstrating familiarity with the literature

Beyond the purpose of literature citations in the documentation of what you write, a sometimes-neglected aspect is the role of these citations in demonstrating your familiarity with your research area. The work you cite is a reflection of your command of the information in your topic, and it is therefore important that you be thorough in acknowledging all relevant publications. While it may not be crucial to cite widely known facts (e.g., HIV/AIDS has killed millions of persons worldwide), citing primary data that support your hypotheses or experimental approach is very important. This shows that you are well versed in all the relevant information and cognizant of all pertinent aspects. While your reasoning seems crystal clear to yourself, it may not be so clear to the reader unless you lead him/her through all the information that led you to your conclusions.

Acknowledging the work of others

Another less direct impact of your literature citations is giving credit to others in the field. Although one citation for a key fact may be enough to provide documentation, failing to provide other relevant citations may leave a negative impression. Omission of those citations can suggest to the reader that you are not thorough in your knowledge of others' work in your field. This also may give the impression that you do not properly consider all the available data. Keep in mind that your reviewers in fact may be among those others working in this area.

Common errors

Incomplete references

All facts in the chain of logic leading to your hypotheses should be supported by citations where possible. Remember that facts that seem obvious to you may be less obvious to a reviewer who does not work in the same area. Also, crediting all groups that have published on a topic will demonstrate the breadth of your familiarity.

Failure to reference alternative viewpoints

While providing citations for the viewpoint supporting your hypotheses is crucial, it is usually important to acknowledge alternative viewpoints or hypotheses and to reference these. If you do not, the reviewers may perceive that you do not know the field well, or that you are not objective in your analysis. Furthermore, it is possible that a reviewer may have an opposing viewpoint, in which case you may head off criticism by acknowledging the other viewpoint.

Misuse of references

Reviewers often check the accuracy of the references, and they may already know the literature well. Errors in your citations, or worse yet, misinterpreting a citation to support your point, may leave a poor impression.

Overuse of citations

While being thorough is important, it may be unnecessarily cumbersome to provide too many references for details that are not central to your project. Because clarity is the overriding issue, having numerous extraneous references may detract by making it difficult for the reader to find the relevant citations.

12

Use of Appendices

Supplemental and not primary material

Appendices are *supplementary* material for your grant application. The application should stand alone, with the appendices being available as supporting documentation if a reviewer wishes to check a point or get further details. They should not be used to convey crucial information that is essential to the project; appendices are provided for the convenience of the reader, not for the convenience of the writer. Technically, the reviewers are not obligated to read the appendices. Also keep in mind that only the primary and secondary reviewers and reader receive them; other members of the study section do not.

Giving peripheral details for the convenience of the reviewers

The proper use of the appendices can enhance the brevity and clarity of the "Preliminary Results." Appended publications should enhance the "Preliminary Results" section by giving peripheral details relevant to the proposed work, but not directly necessary in the main body of the grant. You should not rely on appendices to be the sole documentation for novel ideas or techniques. The "Preliminary Results" section should be comprehensible as a self-contained section, and appendices should serve only to provide optional complementary or peripheral details. Appended papers should be provided to assist the reviewers if they have questions about minor details.

For example, if you have devised a completely new method and propose to utilize it to address an aim in the application, showing results from the method in the "Preliminary Results" section is essential. Not showing the method and simply referring to your publication in an appendix would force the reader to dig through the appendix to learn about this technique, making the application difficult to read. In showing data derived from your new method to demonstrate its feasibility, you may decide against describing the protocol in every last detail, but give only enough description to demonstrate the key principles. Citing an attached appendix gives the reader a resource to find the technical details if necessary; but the data you show in the "Preliminary Results" section should already suffice to convince the reader that the method is working in your laboratory.

Documenting technical capabilities

An appendix may also be useful to demonstrate that you are capable of techniques that are straightforward and established. For methods that are fairly standard, for example, PCR-based sequencing of a virus from clinical samples, you may rely entirely on an appended paper to show that you are familiar with the technique and have used it successfully in the past. The more common the technique, the more acceptable it is to show results without detailed explanation of the methods in the "Preliminary Results" section, and refer to an appendix.

Common errors

Using appendices to circumvent page limitations

The materials provided in the appendices are for the purpose of supplementing the application, and not filling major gaps. If a piece of information is crucial to your application, it should be provided in the "Preliminary Results" or "Background and Significance" sections as appropriate. The correct approach is to show relevant data from your publication in the "Preliminary Results" section, and refer to an appended copy of the paper for further details.

Appending unpublished manuscripts

NIH rules state that appended manuscripts should be only those that are published or in press. Technically, submitted manuscripts are not allowed. It is preferable to show data from such manuscripts in the "Background and Significance" section and mention that the manuscript is under review. However, this rule is sometimes bent, and you should check with your SRA if you wish to append a manuscript that is unpublished.

Appending the work of others

The purpose of the appendices is to show YOUR capabilities and ideas. They should therefore be work performed by you and/or other significant personnel in the grant application. Providing a publication that is not yours (or a collaborator's) contributes nothing toward this goal. It is implicit that appendices are a demonstration of your work (or that of close collaborators).

Irrelevant appendices

The appendices are included to support your hypotheses or provide evidence for technical capabilities pertaining to the project at hand. While including an appendix that does neither probably won't hurt, it is quite unlikely to help your case. NIH suggests that appendices are not intended to be a "dumping ground" for information.

13

Administrative Sections

Importance for acceptance of the grant application

The administrative sections of your application are absolutely required for the grant application to be accepted for review and funding, and errors in these sections are a common cause of delays and rejection of applications without review. It is thus critical that you give the proper attention to completing these sections according to NIH protocol. Familiarizing yourself with the requirements listed in the instructions for writing grant applications should be one of the first steps you take. Full descriptions and instructions are included in the Form PHS 398 documentation, available through the NIH website. Most information in these sections is mandated by federal law, and adhering to these requirements is therefore essential.

As mentioned earlier, some of this information will need to be reviewed by the administration at your institution before the application can be submitted to NIH. It is advantageous to complete first the administrative sections required by your institution, generally the *Face Page* (including approval certification numbers for human subjects and animals), *Description*, *Personnel*, *Budget*, and *Checklist*. The application can then be preapproved by your administration and ready to submit as soon as you finish the scientific portion.

The instructions for each administrative section vary for different grants, so refer to the specific instructions for your application.

Specific Administrative Components

Face page

This is the cover for the application that contains all the important identifying information for administrators at your institution and NIH, as well as for the federal government. Most of these items are self-explanatory and specified in the instructions. Some points of note:

> The title of your application should be short (56 characters including spaces) and descriptive. Strike a balance between titles that are too broad or too specific, and avoid the use of nonstandard terms or abbreviations.
>
> You will need to indicate whether your application is a response to an RFA ("Request for Applications") or PA ("Program Announcement"), so that the CSR can send it to the appropriate reviewers.
>
> It is definitely to your advantage to declare yourself as a "New Investigator" if you qualify, because the reviewers are asked to treat applications from new investigators with greater leniency (especially concerning preliminary data). NIH defines new investigators as those who have not previously obtained major NIH funding. Investigators who have had minor NIH grants (i.e., R03, R15, R21, any K grant except the K02 and K04) still qualify as new investigators.
>
> You will need to provide human subjects and animal welfare approval numbers from your institution, or proof that such approval is pending review. NIH will not release funds to you until these approvals are finalized.

Description

This is a summary of your project that should be written to be understandable to scientists outside your general research area. It should be as clear and simple as possible, so that administrators have an idea of how to classify your application, and assign it to reviewers with appropriate expertise.

Biosketch

This is the standard format of a curriculum vitae for NIH. You should observe NIH instructions closely, including the limit of four pages. The section on research support should include a sentence or two summarizing each project, so that the reader has an idea of your research areas. Publications should include only works that are already published or in press, not submitted or in preparation. The reviewers will use these data to judge your credentials for the research proposal. Also, each collaborator should provide a biosketch. Ensure that all biosketches are up-to-date; otherwise a gap in recent information may be mistaken for lack of productivity.

Personnel

A listing of relevant personnel should be given. These include persons whose involvement and expertise are important to the success of the project, whether or not they would receive funding support from the grant.

Resources and Environment

This section is a listing of the resources available for your research project. You should describe laboratory space, major equipment, accessibility of core facilities, office space, computers, administrative support that are at your disposal. This information contributes to the "Environment" criterion of the reviewers (see Chapter 15).

Consortium Agreement

This is the agreement between your institution and another institution when funds from your grant are provided to a coinvestigator outside your institution. NIH requires that a formal signed agreement be provided, to ensure that administrators at both institutions understand the terms of the fund allocation.

Budget and Budget Justification

Guidelines vary for the type of grant, and are given in the specific instructions for each grant application. Funds are requested in terms of "direct costs," which is the net amount directly available to the investigator if the grant is approved. "Indirect costs" (also known as "overhead") include additional money (a fixed percentage of direct costs) that is allocated to the institution to cover infrastructure costs associated with research. The rate of overhead differs according to the institution, which has a negotiated agreement with NIH.

For most independent researcher grants, the budget must be "modular" if under $250,000 (direct costs) a year, and "detailed" if more. Detailed budgets must include a precise breakdown of anticipated expenditures; this is usually applicable to large program project grants or large R01 grants to established investigators. Newer investigators will usually utilize a modular budget. Modular budgets are requested amounts in increments of $25,000 "modules," up to 10 modules. For these budgets, the only justification detail you must provide is that for salary support of staff. In the "Budget Justification" section, you must list all persons who will contribute to the project, including summaries of their expertise, role, and the "percent effort" (the percentage of their overall research time and effort) that will be applied to the project. Important persons who will assist but receive no salary support should be listed also, and designated as "no salary support requested."

Applications requesting large budgets (more than $500,000 in direct costs for any single year of the grant) require preapproval from the receiving institute before submission. For such applications, the responsible program officer should be contacted to arrange this preapproval; otherwise CSR will reject the application without review.

Human Subjects

This is required for any research (including purely observational studies) that examines humans or utilizes materials from humans (such a peripheral blood), and inadequate documentation will result in rejection of your application without scientific review. In addition to the institutional assurance documentation, you must have an additional descriptive section on the human subjects who will participate in your research. This is a separate subsection that is actually a required part of the scientific section of your application (even if the research involves no human subjects, in which case this must be clearly specified). Read the NIH instructions carefully. For research involving human subjects that is not considered a clinical trial, you must address the following categories:

> *Protection of human subjects.* Risks involved to participants, procedures to minimize risk, informed consent procedures, procedures to maintain confidentiality, and other relevant points.
> *Inclusion of women and minorities.* Inclusion criteria for research subjects, whether women and ethnic minorities will be specifically chosen or excluded, and how it will be ensured that the results are applicable to women and minorities if applicable.
> *Inclusion of children.* Whether children will be included or excluded, and criteria for these decisions.
> *Ethnic/racial targeted/planned enrollment table.* A table provided by NIH that breaks down study participants by race and gender, to indicate the targeted enrollment of persons for your study.

Each of these should be given an individual subheading and described fully, including careful explanation if you are specifically selecting or excluding any population (race, gender, age). There must be clear scientific justification for any bias in the selection of human subjects. State the obvious; for example, a study on prostate cancer will exclude women and children because they are not at risk for this cancer. As a government agency, NIH is mandated by federal law to demonstrate fairness in the expenditure of resources (tax dollars).

If your research does not include human subjects, simply state this under the Human Subjects heading. Some human materials (e.g., commercially purchased blood cells from anonymous donors) are considered exempt from IRB approval, and this should be stated if it is the case for your work. If in doubt, contact your institution's IRB and/or the responsible NIH Program Officer for your application.

Vertebrate Animals

Similarly, NIH has policies to protect animals used for research. If your research involves any vertebrate animals, this is a required portion in your application, which otherwise can be rejected without review. In addition to documenting your institutional approvals for working with animals, you must include a separate heading at the end of your scientific section. Your discussion of animals must address the following points:

> *Description of the proposed use of animals.* Details on species, strains, sex, age, numbers of animals.
> *Justification for the use of animals.* Explanation for the choice of animal species and numbers, and why there is no alternative for the use of these models.
> *Veterinary care for the animals.* Availability of veterinary staff and facilities.
> *Procedures to minimize discomfort/injury.* How these will be limited only to what is scientifically unavoidable, usage of anesthetics, and methods of restraining animals.
> *Procedures for euthanasia.* How euthanasia will be performed if necessary, adherence to the recommendations of the American Veterinary Medical Association.

Checklist

This is a self-explanatory list of requirements for submission of your application.

Cover letter

You should submit your application with a cover letter. This letter will not go to the reviewers, but will be read only by the staff members at CSR that handle your application. You should state the title of your application, and preferences concerning its review. You can request assignment to a particular institute and/or study section (IRG, see Chapter 15). It may also be important to list any persons whom you believe would not be able to review your work objectively, such as scientists with strongly different views, and direct competitors. The CSR will generally honor such requests.

14

Collaborators and Consultants

Adding skills, expertise, or reagents to the project

Your research project may require skills or materials that are not immediately available in your laboratory. In this case, a central question of reviewers will be whether you will have the capability to perform that work. This judgment will depend on your prior experience and track record. An effective way to address this concern, particularly if you are not a highly experienced investigator, is to enlist the aide of collaborators and/or consultants. These should be persons who are easily recognized as experts in a technique that you require, or who possess necessary materials (e.g., access to patient populations, particular monoclonal antibodies, etc).

Defining the role

Collaborators and consultants are defined by their role in your project. Collaborators are persons who directly participate in your work. They may offer to perform an assay in their laboratory, or to train your staff and provide technical backup. Persons who will provide important reagents or samples can also be considered collaborators. Collaborators should indicate clear involvement your research project, usually reflected by a commitment to a percentage of their scientific effort in the budget justification (see Chapter 13). Consultants are persons who are available to give you advice. They generally are not directly involved, but should indicate their availability to provide you with guidance when needed. These are usually widely recognized experts in their field, whose input would provide intellectual support for your work. In contrast to collaborators and consultants, co-primary investigators are persons with whom you share the responsibility and credit for the project.

Letters of support

Letters of support should be provided by both collaborators and consultants, and these should contain identification of the collaborator/consultant, their expertise, and the role they will play in your work (doing assays, providing technical advice, providing reagents, training your personnel, etc). This provides clear documentation of how they will assist you. They should refer directly to your grant application by title, and be as specific as possible. These letters are included in the appendix.

Maintaining an independent role

While collaborators and consultants can bolster your application, care must be taken to preserve your role as the lead researcher if you are applying as an independent investigator (for R series grants). The role of expert collaborators and consultants should be defined clearly to avoid detracting from your independence as innovator of the project.

Common errors

Missing letters

Any consultant or collaborator should provide a letter documenting his/her role in your project. Surprisingly, such letters are often missing from applications, in which case the input of that person is questionable to reviewers.

Vague letters

A letter lacking specific details can make the support for your project unclear. For example, statement such as "I fully support your project on IAPCN" is far less valuable than one stating "My laboratory will provide your group with 20 IAPCN mice and 10 negative control mice per month for your studies."

Lack of sufficient effort

For collaborators with a substantial role, the effort they allocate to your project (percent effort) should be adequate to reflect their commitment to the work. For example, the involvement of a collaborator whom you state will perform and analyze thousands of assays for your project, but who lists only 1% effort, is likely to be criticized for lack of commitment.

15

Scoring Process

Processing of the application by the CSR

Once your grant application has been submitted to NIH, it undergoes a standard process. Administrative officials of the NIH Center for Scientific Review (CSR) first check the application for completeness of administrative data and conformity to format requirements. Applications that are incomplete or violate rules (e.g., page limitations) are usually rejected without review. If the application is a response to an RFA or PA, the appropriate program officer will review it for relevance to the RFA or PA; an application deemed inappropriate can be reassigned outside the RFA or PA as an unsolicited application, or outright rejected without review.

The application is designated with a unique identification number (specific for the institute, e.g., AI043203 for an application to the National Institute of *A*llergy and *I*nfectious Diseases), which you should use when contacting NIH. CSR then assigns the application to an institute and "Integrated Review Group" (IRG), or "study section," which is a panel of outside experts assembled to review grants. This is a peer review process, because these scientists are from the scientific community at large. The study section convenes and assigns the grants "priority scores" that are a judgment of overall scientific merit, and a written review and score are sent to you within a few weeks. Funding will then be decided at the institute based on these scores.

For R01 applications, there is an additional internal review by a "Council" meeting of external scientific advisers at the assigned institute. The council makes the final funding decisions, largely based on the priority score for each grant application and the recommendation of the institute program officials.

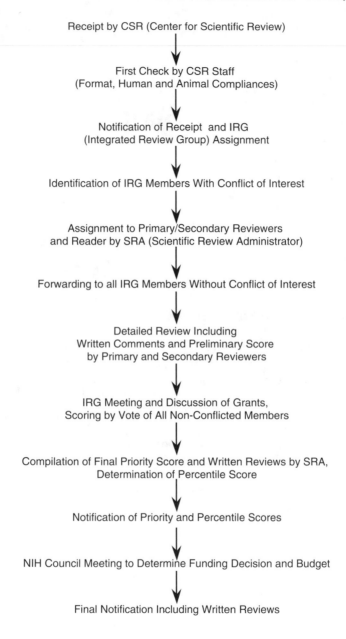

Receipt by CSR (Center for Scientific Review)

↓

First Check by CSR Staff
(Format, Human and Animal Compliances)

↓

Notification of Receipt and IRG
(Integrated Review Group) Assignment

↓

Identification of IRG Members With Conflict of Interest

↓

Assignment to Primary/Secondary Reviewers
and Reader by SRA (Scientific Review Administrator)

↓

Forwarding to all IRG Members Without Conflict of Interest

↓

Detailed Review Including
Written Comments and Preliminary Score
by Primary and Secondary Reviewers

↓

IRG Meeting and Discussion of Grants,
Scoring by Vote of All Non-Conflicted Members

↓

Compilation of Final Priority Score and Written Reviews by SRA,
Determination of Percentile Score

↓

Notification of Priority and Percentile Scores

↓

NIH Council Meeting to Determine Funding Decision and Budget

↓

Final Notification Including Written Reviews

Grant assignments and the study section (IRG)

The study section is a panel of scientists with defined areas of expertise recruited by NIH to review grants. Each section has a particular thematic focus (e.g., for HIV/AIDS, basic virology, pathogenesis, or AIDS-associated opportunistic infections), and your grant is assigned to a section with relevant expertise. If you do not request a specific section, the scientific review administrator (SRA) decides the assignment based upon the topic of your application. It is usually to your advantage to request a particular section, since you can make a choice based on the interests and expertise of the members. Applications responding to an RFA or PA are usually reviewed together in the same section.

The applications are given to the reviewers several weeks in advance of the study section meeting to assign scores. The "scientific review administrator" (SRA, the CSR administrative official in charge of organizing the section), assigns the applications to the scientific reviewers in that section, based on their specific expertise and research interests. Each application has three reviewers with the main responsibility: the primary and secondary reviewers, and the reader. These three reviewers read the application in detail, and are responsible for deciding preliminary scores before the study section meeting. The primary and secondary reviewers also produce drafts of written critiques. The other scientists in the section also have the opportunity to read all the applications, although each person's major focus will be the multiple applications for which he or she serves as primary or secondary reviewer, or reader.

NIH issues formal priorities to guide the reviewers in the scoring of grant applications. The written critiques provided by the primary and secondary reviewers discuss the strengths and weaknesses of your grant application in several categories deemed important in the assessment of your research project. The specific areas and the emphases vary depending on the type of grant, but the two main classes are those for independent investigators and training grants.

Scoring criteria for independent investigator grants

The scoring criteria for independent investigator grants (such as the R series) include:

Significance. This is the importance of the research questions being posed. The reviewers consider whether the work in the proposal would have an impact on the field, through advancing the state of knowledge, changing a paradigm, or opening a new area. For applications responding to an RFA or PA, the applicability of the proposed work to the RFA or PA focus is a key criterion here.

Approach. This is the quality of the experimental planning. The research plan is assessed for feasibility, soundness, and logic. In the written evaluation, the reviewers summarize the approach and dissect your research design, usually aim-by-aim. This review section usually contains the most detailed comments and critique, because the experimental plan is the heart of the application.

Innovation. Here the reviewers comment on the overall novelty of the application. The innovative qualities of the hypotheses and experimental approaches are weighed, and originality of the ideas and methods is assessed. Be aware that novelty is a double-edged sword; it can have a negative impact if one does not have the data and experience to demonstrate feasibility. The more innovative the application, the more strength is required in its justification.

Investigator. The primary investigator's credentials and those of other key personnel are assessed. The reviewers consider whether the prior training and productivity are adequate for the project. For a new investigator, a key factor is the track record of independent research. It is therefore important to demonstrate independence of thought and decision-making, and to make it clear that the ideas and plans in the application are entirely original. Collaborators are also considered in this section. These assessments are based on information in the biosketch(es).

Environment. Here the reviewers assess the research environment and resources available for the proposed work. The infrastructure of the institution(s) where the grant would be funded and its adequacy for the project is evaluated, based on the information provided in the "Resources and Environment" section. The other key consideration is the degree of support being provided by the institution to you as an independent investigator, which may be conveyed in a letter of support. Important points include the academic appointment and career path offered, the resources (e.g., core facilities, laboratory space, funding) available, and the general commitment of the institution to the success of the applicant as an independent researcher.

The relative weight of these criteria is left to the personal judgment of the reviewers. An application need not be outstanding in all areas to receive an outstanding score, and almost none are. For example, a proposal may describe straightforward approaches that are not novel, but which will yield an important advance of high significance. The final decision is a subjective combination of all criteria, dependent on the type of grant. Grants for larger projects such as R01s are generally weighed for the likely results and success, while smaller grants targeted for high-risk projects such as R21s are slanted to favor innovation. NIH stresses that the scoring is not based on comparisons to other applications, but to an idealized standard.

Scoring criteria for training grants

The scoring criteria for training grants (such as the F and K series) include:

Candidate. The applicant's credentials are considered in relationship to the proposed project for further growth as a scientist. In contrast to the "Investigator" category above, this section is focused on the potential to develop into an independent investigator, rather than the track record of prior achievements. The recommendation letters contribute a major portion of data for this assessment, in addition to the educational history given in the biosketch.

Sponsor and training environment. Here the credentials of the mentor and his/her prior record of training new investigators are assessed. The central issue is whether the mentorship and laboratory facilities of the sponsor will provide an environment conducive to training the candidate to be a successful independent investigator.

Research proposal. This section is an abbreviated version of the "Approach" section for independent investigator grant applications, also incorporating the "Significance" and "Innovation." Because the training candidate is not an established investigator, the review of the research plans is less detailed and mostly concerned with the overall soundness of the research. For these applications, the research is considered a vehicle for training of the candidate, rather than the primary objective of the grant. As such, the impact and novelty of the work are not as crucial as demonstrating that the research project offers the applicant a rigorous exercise in planning and executing a solid project that will provide meaningful results.

Training potential. This section is an assessment of the overall training plan. The key question is whether the program as a whole will be useful for training the candidate to be an independent investigator. The mentor's descriptions of the formal activities for the trainee (classroom courses, lectures, lab meetings, national meetings) play a key role here.

Study section meeting

The study section meeting is usually held in the Washington DC area. At this meeting, the reviewers assemble for a verbal discussion of all the applications to determine their final scores. The applications are discussed one at a time by this panel. The discussion is opened by the primary reviewer, who introduces the application and describes the research, going through the aims and experiments in enough detail to give everyone a general understanding. The primary reviewer then gives his/her assessment of the work, including a score. The secondary reviewer and reader then follow with their assessments and scores, including comments about whether they agree or disagree with the other reviewers. The floor is then open for questions and discussion. If the scores of the three reviewers differ widely, there is debate. The other reviewers in the section skim through the application and ask questions or clarify points raised by the primary, secondary, and reader. After this discussion process, the primary, secondary, and reader have the opportunity to revise their opinions and scores (and modify written critiques). Discussion/debate continues until opinions are solidified, and then each member of the study section records a written vote for a priority score. These numbers are subsequently averaged by the SRA to obtain a final score from the entire panel. The SRA also provides a written summary of the discussion for the applicant ("Resume and Summary of Discussion").

The "priority score" and "streamlined" scoring procedure

"Priority scores" are intended as pure reflections of the scientific merit of the application (based on the above criteria), and are expressed on a scale of 100–500. A score of 100 is best, and 500 worst. The numeric priority score assigned to your application is a composite of all scores given by the study section members after the discussion described above.

For some types of grants (e.g., the R01), the discussion at the meeting may be "streamlined" and the application given a score of "lower half" or "unscored." This indicates that the major reviewers (primary, secondary, and reader) all agreed that the application was scored in a range that clearly would not earn funding (usually 250 or higher). This process is intended to reduce the time spent during the meeting for discussion of grants that will not be funded. Streamlined grant applications, while not fundable, are provided with feedback in the form of the written comments of the primary and secondary reviewers, just as the applications that receive scores. These reviews serve as feedback to allow the applicant the opportunity to revise the application after responding to criticism. Training and developmental grant applications are not streamlined.

The reviewers and conflicts of interest

Study section panelists are chosen by NIH as experts in their areas of research. These individuals are independent investigators who themselves have a successful track record of NIH funding, who administrators feel would serve as objective critics. Reviewers must formally agree to ethical standards including preserving the confidentiality of the information in the grant applications. They receive a small per diem stipend and reimbursement of their travel expenses for the time spent at the study section meeting, but essentially donate their time preparing for the meeting (detailed analysis of several applications, and providing written critiques) as a voluntary service to NIH.

Before the study section meeting, all reviewers in the section receive a listing of all grant applications to be reviewed at the meeting. They are then each required to identify any applications that might pose a "conflict of interest," real or perceived. NIH rules are strictly defined, and designed to exclude the input of any reviewer whose objectivity could be questioned. Conflicts include: having collaborations (past or present) with the applicant, belonging to the same institution as the applicant (within the past 3 years), training or being trained by the applicant, having personal ties to the applicant, being in direct competition with the applicant, having financial interests in the project, or any other factor that could be perceived as a conflict (whether or not an actual conflict exists). Reviewers do not receive a copy of applications with which they are "conflicted," and are required to leave the room when they are discussed at the study section meeting, effectively excluding them from any scoring input or knowledge of the discussion.

Administrative sections

The administrative sections (Budget, Human Subjects, Animals) do not contribute to the score given by the reviewers, which reflects only the scientific merit of the project. However, if the reviewers find serious ethical concerns concerning human or animal participation in the research, the grant can be rejected without a score. As part of their assessment, the reviewers are asked to provide comments concerning the use of human subjects and/or animals in the research, and the budget. They review the documentation provided in the grant application concerning human participants (institutional review board approvals for human subjects, inclusion of racial minorities/women/children) and animals (institutional animal ethics committee approvals) for adequacy, and assess whether the budget presented in the application is justified for the work proposed. They may make budget reduction recommendations if they consider the budget too high (to cut the amount and/or time of the award). These comments are mainly for the use of the scientific administrative staff at NIH, who will make the final decisions concerning these issues, and do not weigh into the score.

The "percentile score" and funding decision

Whether a grant is funded depends next on the administration at NIH. After the priority score is assigned, most applications (e.g., R01) get a "percentile score" in comparison to the priority scores given to all other grants reviewed by the same study section over the past year. The percentile score is the final figure used to determine whether or not the grant is funded. This system helps correct for any scoring bias of the particular study section, because different sections may be more or less lenient in their assignment of priority scores. Depending on the specific research topic and government funding levels for that institute, a percentile score cutoff is established to determine the applications funded and rejected. In unusual circumstances, the program officers can choose to fund an application that does not meet the cutoff, based on the priorities of the institute and availability of additional funds.

Some applications (e.g., an R21 response to an RFA or PA) do not receive percentile scores. These applications are funded based on the priority score relative to all other applications in the same pool. NIH often has fixed funding for the RFA or PA, in which case grants will be awarded to the highest scoring applications until the funding is entirely utilized. Thus, depending on the amount of money and number of applications, awards could range anywhere from one to all applications responding to the RFA or PA.

16

Resubmitting an Application

The majority of grant applications are not funded, due to competition for limited funds. It is common to resubmit after initial rejection. Applications are allowed three chances for review and funding (with the exception of RFA/PA responses, which may be time-limited). Each applicant receives the written critiques of the primary and secondary reviewers (and occasionally the reader or other study section members who wish to contribute comments), plus the overall summary of the study section discussion from the study section senior research administrator (SRA), in the "Resume and Summary of Discussion."

Analyzing the critiques

Resubmission of the application requires revision in response to the critiques from the prior review. These critiques should be analyzed thoroughly, to understand fully the problems perceived by the reviewers. In reading these comments, you should keep an open mind and avoid feeling personally slighted; objectivity and clear thinking are essential in your response. After familiarizing yourself with the critiques, begin with an analysis of common themes in the reviews. Criticisms shared by both primary and secondary reviewers are likely to indicate significant problems that require definitive correction. Likewise, recurring themes in the reviews probably point to deficiencies that need to be addressed. Finally, you should consult your program officer for guidance, because he/she often has attended the study section meeting and may have further insight not clear from the written critiques.

Interpreting the score

Usually the score is a clue as to how extensively you should make modifications. A score that almost earned funding suggests that your application is basically sound, and probably needs only minor modifications. A score in the "lower half" tends to signify that the reviewers felt there were significant fundamental flaws, requiring a major overhaul. In this case, they most likely felt that the reasoning behind the hypotheses or approach had serious errors that require major corrections.

Whether to change your aims

The first major decision regarding the resubmission is whether the general struc-ture of your application needs modification, that is, whether the hypothesis and specific aims require alteration. This depends on the nature of the criticisms. If there is clear agreement that an aim is seriously flawed, that aim may need to be extensively changed or entirely removed. If the reviews consistently identify a gap in your work that needs filling, it may again require extensive change of an aim or even addition of a new aim. A third reason to modify the aims is to clarify the organization of the application as a whole, in response to the criticism that it lacks focus and clarity. In this case, you may need to reorganize your ideas into new aims that change the emphasis of your project.

Modifying the experimental plans

The next decisions involve modifications of experimental plans to address weak-nesses or errors identified in the reviews. Oftentimes the reviewers may ask for the addition of controls, or provision of more experimental details, and these are straightforward to address. It may be necessary to provide more information and/or preliminary data if a procedure is questioned, or perhaps an alternative approach needs to be added or substituted. Sometimes a criticism may be a matter of misunderstanding, in which case further explanation but no actual modification is required. Also, you may need to adjust the introductory and concluding sections to modify your reasoning or its explanation, based on the reviewers' disagree-ments or misinterpretations. Depending on the nature and vigor of the criticism, you appease the criticism by simply acknowledging caveats, or you may need to alter your reasoning if it contains fundamental flaws.

Deciding the focus of your changes

When considering changes to make, it is important to assess the components of actual deficiencies in your reasoning and/or plans versus deficiency in their communication resulting in the criticism. This usually requires some in-depth analysis of the comments. It may be difficult to know how much of each factor is to blame for causing a criticism; the key is to understand the critiques as fully as possible. Also, the revised application does not necessarily return to the same reviewers, although it usually goes to the same section again. This can complicate the task of revising the application, because the new reviewers may not necessarily have the same viewpoints as the prior reviewers. Thus you need to prioritize making adjustments to satisfy the prior reviewers while not opening the revised application to new criticisms. The comments of the prior reviewers are not binding to the new reviewers, and so you must apply your own judgment as to whether and how you want to implement changes in response to the criticisms.

In other words, simply modifying your application to address literally all the comments in the reviews does not guarantee an improved score for several reasons:

1. The new reviewers may not have the same opinion.
2. Your modifications may have new flaws that can be criticized.
3. The criticisms of the prior reviewers may have been intended as general suggestions rather than literal instructions.
4. The criticisms of the prior reviewers may have been based on misunderstanding of your presentation rather than faults in your plan.
5. Your modifications may not fit the rest of the project, making it seem patched rather than changed.

There are therefore no simple rules for responding to the critiques. The ultimate goal is to produce an improved project, and this is not necessarily the same as mechanically responding to the comments. In the end, you must analyze the critiques, understand the reasons for the specific criticisms, and address the *causes* of the criticisms rather than reflexively responding in a literal manner. Showing your ability to think critically about your work is more important than blindly trying to appease the reviewers. Again, the program officer may have additional insights, and can be a key source of guidance for your response.

Highlighting revisions

To facilitate the re-review process, any changes since the last application should be indicated in the new document. This can be done by putting revised text in bold font, or by shading using a revisions function in your word processor. The new reviewers (whether the same or different persons from the prior review) do receive copies of your prior review, and will be aware of the prior criticisms. They will check the grant for your responses to these specific comments, point-by-point. Highlighting changes therefore assists them in assessing your responsiveness to the prior comments.

Writing the "Introduction to the Revised Application"

Resubmitted applications are allowed this additional section to respond directly to the comments of the reviewers. For most applications, this is three extra pages to introduce the revised scientific section. The purpose of this section is to provide an overview of the changes you have made in response to the previous critiques. It should begin with a paragraph broadly describing the major revisions and significant changes in your thinking since the last submission. The specific comments of each reviewer should then be individually addressed, point-by-point, like a manuscript resubmission cover letter. It is best to restate each reviewer comment as a direct quote, followed by your response, including a description of how/where the manuscript has been altered. Finally, there should be a brief summary statement. To continue our fictional example of a grant application to study IAPCN, an "Introduction to the Revised Application" could begin:

> We thank the reviewers for their constructive criticism and insightful comments of our first submission of a grant application to study IAPCN pathogenesis. In response, we have modified and completely rewritten several sections, which are indicated in bold font. Generally, both reviewers found Aim 2 problematic in its lack of detail concerning specificity-controls for the small molecule inhibitors and enhancers of the NEC pathway, several of which have been shown to have additional effects on serotonin reuptake in the brain. This raised concerns that the planned experiments would not be able to resolve the causality of NEC versus serotonergic alterations induced by these molecules. We have therefore added a new section describing additional control experiments designed to address this issue, in the "Research Design and Methods" for Aim 2 (page 21 of the revised application). Also, both reviewers criticized Aim 3 for lack of documentation of human subject sources, and felt that we did not show access to enough subjects for the proposed studies. We have therefore added a letter of collaboration from Dr. B. R. Ain, head of the IAPCN clinic at our university medical center, documenting a recruitment pool of approximately 700 patients, and included power calculations in the revised Aim 3 (page 23).

> *Specific Responses to Primary Reviewer*
> "Figure 3, demonstrates that the NEC factor NKF-245 is significantly elevated in IAPCN mice compared to normal control mice. However, this is apparently contradicted by the data in Figure 6, where NKF-245 expression is undetectable in IAPCN mice."
>   We did not intend to suggest that NKF-245 levels are static in the course of IAPCN, and should have discussed our preliminary findings on changing NEC levels over time. We now address the apparently contradictory results by adding a new figure to the Preliminary Data showing a time course experiment, where NKF-245 is initially elevated early in the course of IAPCN, and then depressed (Figure 4 in the revised application). Describing the kinetics of all NEC factors is now an additional section in Aim 1 (page 19) ....

Common errors

Misinterpretation of critiques

Often the written critiques of your application are reasonably straightforward, but occasionally it may be difficult to prioritize the criticisms or dissect their underlying reasons, especially if the comments of two reviewers are widely divergent. In such cases, it may be useful to consult the Program Officer at NIH in charge of your grant. This official often attends the study section meeting to listen to the verbal discussion of the application, and can sometimes provide insight that is not apparent from your written review.

Being overly defensive in the "Introduction to the Revised Application"

This section should not be treated as a rebuttal, and you should view the criticisms as constructive (even if they are negative). Some applicants become extremely defensive and lose their objectivity. Occasionally, some frustrated applicants have resorted to questioning the knowledge of the reviewers, attacking them for misunderstanding a point, or even outright launching insults at them. This attitude is counterproductive to say the least. You should keep in mind that any criticism offered by a reviewer reflects either a true flaw in the application, a misunderstanding on the part of the reviewer, or a difference in opinion on the part of the reviewer. As stated above, it is crucial to analyze the comments to find their underlying reason, and to address that cause. Scientific flaws should be corrected. Misunderstandings on the part of the reviewer should be clarified, bearing in mind that it is YOUR responsibility to make the ideas clear. Remember that the reviewers are knowledgeable in their own area, but may not be experts in yours. Differences in opinion should be handled respectfully, by acknowledging the possibility of alternative viewpoints, providing evidence for your viewpoint, and discussing how the research plan will address these differences.

Not being responsive to the critiques

A frequent criticism during reviews of revised applications is "lack of responsiveness" of the revision. The root of this problem is lack of change in areas that the reviewers consider significantly deficient or flawed. This usually occurs when the applicant either argues about a criticism without modifying the plan, or simply ignores a criticism. Again, understanding the underlying reasons for the criticisms is the key to avoiding this problem. If you choose not to alter an approach in response to a criticism, you must clearly justify the reasoning. The burden is on you to show why no change is needed, not the reviewer to show why it is. Just ignoring a criticism with which you disagree is highly likely to cause difficulties, because the new reviewers (even if different persons) will check the revised application against the prior reviews.

Taking the advice of the reviewers too literally

Sometimes the reviewers will make concrete suggestions (although reviewers are generally instructed not to do this), and this can pose a quandary if a suggestion is out of line with your intent. As per the recurrent theme of this chapter, the key point is to understand the rationale behind the criticisms, and to interpret them properly, rather than responding blindly. Direct suggestions from the reviewer should not be taken literally, because they may be based on incomplete understanding of your project, or based on opinion and style. The ultimate responsibility for improving the application is yours and not theirs, and it is up to you to determine the appropriateness of any suggestions to ensure that any changes fit the scope of your work and improve upon it. In the context of the revised grant, the new reviewers may have differing opinions from the prior reviewer. Again, you need to address the underlying reason for the criticism. Because you are ultimately more familiar with your work than anyone else, you are the best judge of how to address the issue, and you may find a better solution than that suggested by a reviewer. As long as you address the underlying concern and improve your work, it will be to your credit.

17

Submitting a Competing Renewal

Differences from a new grant application

The competing renewal grant application (for renewable grants such as the R01) is treated the same as a new application in most respects. This grant will be reviewed by NIH alongside new grants of the same type, and scored/funded in parallel with those grants. The major difference between a competing renewal and a new submission is in the "Preliminary Results" section, which contains a "Progress Report." The reviewers will assess this section to judge your success in meeting the aims of the previous grant. This provides the opportunity to showcase your progress in the work you proposed in the first grant. As in a new application, the data should represent a cohesive story that sets the backdrop for the proposed research. Again, this section should serve to demonstrate the capability to perform the proposed work, and to experimentally approach the questions being raised. Usually, the new aims in the competing renewal should be a natural outgrowth from the results garnered during the previous funding period.

Reporting progress from the previous funding period

A brief synopsis should be provided to introduce the work performed during the prior funding period, including an explanation of how that work fared in achieving the past stated specific aims. A subsection should then be provided, with the title "Publications Resulting from [Grant #]," listing all relevant publications supported by the previous funded grant. Data then should be presented as in the "Preliminary Results" section of a new application (Chapter 9), under headings of the prior aims. It is important to restate the aims of the prior funded grant as section headings, with data addressing each aim falling in the appropriate subsections. This allows the reviewers to assess the accomplishments in addressing the prior goals of the project. Care should be taken to provide a clear logical link between the questions answered by the data (old aims) and the questions raised by the data (new aims). Ideally, the work accomplished during the prior funding period should provide answers concerning the prior questions and hypotheses, and open new avenues of research that are addressed by the new aims. If prior goals are not met, or prior hypotheses are disproved, this is perfectly acceptable if there are reasonable explanations (e.g., unexpected developments in the field, experimental results that moved the research in another important and relevant direction). In the end, the results and process should be clearly justified, whether or not the work deviated from the original aims. The results should be related to the original aims, to give the reviewers a sense of how the work contributed to advancing the field, whether or not you modified the plans.

The reviewers will have access to the prior reviews from the earlier grant and will be aware of the prior aims and criticisms. Providing proof of strong progress toward the work originally proposed, and capability to deal with problems and respond with alternative approaches can be compelling evidence for your success in the new application. Because the ability of an investigator to analyze and deal with research obstacles is an asset that receives heavy priority in grant scoring, this is an area in which you can capitalize by laying out your progress in a manner that impresses the reviewers. For example, if you have successfully invented a technique such as protein-to-RNA reverse translation as an approach in the prior funding period, your credibility in proposing to develop a new technique such as *in situ* reverse translation in your new application will be viewed with far more optimism than if you proposed this technique with no prior track record.

Scoring of the competing renewal application

The final judgment for the competing renewal application will be based on its relative merit compared to other applications in the same review section, just like any new application. The advantage of submitting the competing renewal is thus the opportunity to show your track record and ability to carry through a proposed project. A record of productivity toward the previous aims, and resulting multiple publications in reputable journals is a key advantage over submitting an entirely new grant application. Conversely, lack of productivity and publications can be an extremely negative factor in the scoring of a competing renewal, putting it at a disadvantage.

Common errors

Failure to address prior specific aims

The reviewers have access to the prior reviews of the funded grant, and will therefore have a listing of the prior specific aims. If progress toward any of these aims was poor, not listing the aims in the competing renewal will not hide this fact. It is far better to explain why progress was not achieved than to hope that the reviewers will overlook the omission.

Presentation of work that is not relevant

Some writers attempt to inflate the productivity on the past grant by including related research in their laboratory, in the form of data and publications from other projects. Usually, such "filler" is obvious and leaves a poor impression. Reviewers will check the information against the prior aims, and be aware of your other projects from reading your biosketch. Other work you have accomplished, no matter how interesting, will not be to your credit on this project if it does not fall under the aims. Of course, other work that is relevant to a new research direction in the competing renewal should be given as preliminary data or cited, but it should not be presented as progress in the prior project.

Ignoring prior aims

It is unavoidable that unforeseen circumstances may preclude success in an aim. Sometimes it becomes clear that the aim is unachievable, other times the aim may be rendered unnecessary by progress in the field. In all cases, it is important to explain why the aim was not achieved. If, however, an incomplete aim addresses a relevant question that can still be answered, it is unwise to ignore this aim and move on to something else. NIH places priority on results from funded grants, and commitment to the stated project is important. You want to avoid the impression that you do not fulfill your responsibilities for funded projects.

Repeating prior aims

Similarly, if aims in the prior application were not completed, restating these aims as the basis for a renewal generally does not get a favorable review. Lack of progress on an aim should be clearly explained, and if the aim will be reused for the competing renewal, there should be valid justification for why the aim was not attained, and explanation of innovative developments that now enhance the probability of success. This explanation should contrast the new approach with the prior approach, to avoid the appearance that the new approach is simply a recycling of the previous attempt.

Disregard for prior reviews

Disregard for the prior critiques, leading to problems in the work is a common situation that casts doubt on the responsiveness of the investigator to criticism. Even after your grant is funded, it is wise to be familiar with the critiques of the work to ensure that you avoid any foreseeable pitfalls.

Submitting the renewal late

There are several months of lag time between review of a grant and the start date for funding. To be safe, it is wisest to submit the competitive renewal as early as possible, to allow yourself time for resubmission if it is not funded.

18

Non-NIH Grants

The principles of writing an effective grant application for the NIH are universal, and are generally applicable to applications to other organizations. In fact, many foundations and other granting institutions adopt the NIH format exactly or with minor modifications, and organize review panels that score applications in the same manner as NIH. When applying for these grants, it is crucial to be familiar with the instructions concerning format, style, administrative requirements, and other details of the application. Despite any differences from NIH, however, the strategies and goals in writing are usually the same.

The overriding target of your application remains efficient communication with the reader. As with NIH applications, tightly organized sections, clarity in presentation, and attention to details that affect the reception of your application are goals to have in mind while writing. Most large organizations (e.g., the National Science Foundation, American Cancer Society, Doris Duke Foundation, Howard Hughes Foundation, Gates Foundation, and others) utilize the same peer review system as the NIH, and the considerations of calibrating your writing to your target audience, as discussed for NIH grants, remain applicable. Again, adhering closely to the specified format is important, because the reviewers will be reading your application with this context in mind, and able to absorb your ideas more easily if the information is organized in the standard format. Like an application for an NIH grant, your writing should provide the background for your work, pose questions supported by this background, and provide aims with plans for answering the questions. All organizations, no matter the required format, ask for these key elements.

Just as it is essential to understand the goal of NIH when writing your NIH grant application, it is perhaps even more important to know the goal of other organizations to which you apply for grants. These organizations often have a clearly-defined agenda for spending their money on grants, whether it be encouraging risky new ideas for an AIDS vaccine, training new physician-scientist oncologists, or establishing an infrastructure to treat tuberculosis overseas, and thus your application should be tailored to meet their goals. You should be familiar with the mission of the organization, and its immediate and long-term goals, so that the emphasis of your project is compatible and attractive to their reviewers.

Common differences from NIH grant applications include personal essays delineating your background and interest in the targeted research area, and sections describing the relevance and impact of your project in regards to this area. Unlike NIH, these portions may carry heavy weight in the scoring, because they are directly pertinent to the agenda of the organization, and distinguish its goals from NIH and other organizations. Attention to the reasons for these sections is therefore important. Private foundations and other nongovernment organizations usually have limited resources compared to the NIH, and thus they want to target their research funds to a particular niche. These sections of their applications are often geared toward determining the fit of your project into that niche.

Finally, as with the different types of NIH grant applications, these organizations may offer grants for various goals ranging from training young investigators to funding productive projects from senior investigators. It is again crucial to understand the purpose of the grant, for example, whether the organization seeks to fund high-risk innovative projects, or lower innovation but high-yield projects. Thoroughly understanding the mission of the organization, and speaking to a representative who is their equivalent of an NIH program officer, can offer crucial insights to direct you in writing the application.

19

Conclusions

If you have not seen it enough times in this guide, the key word is *clarity*. In the end, your ideas and plans will not be appreciated if they are not communicated to the reviewers quickly and efficiently. Most of the suggestions in this guide are intended to enhance the clarity with which your application is presented to the reader. Organization and attention to details in your presentation are central to a well-written application.

Another key is *brevity*. The reviewers are busy scientists, and it takes hours to read an application thoroughly. The fewer words it takes to convey your information, the better. You want to keep the attention of the reviewers, make it easy to remember your points, and make it easy to locate information in your application.

Finally, *documentation* is a key aspect. The application is a vehicle to convince NIH that your work is relevant, important, and achievable. You must therefore make your case with facts and hard evidence, that is, published findings and/or data. Being thorough and comprehensive in providing support for your viewpoints is crucial. Providing the information to convince reviewers that you have the ability to perform the proposed work is your responsibility.

Appendix:
Useful Web Resources

NIH provides a plethora of information concerning the application process. Most of this is available on the Internet, including electronic copies of forms and instructions, listings of deadlines, listings of study sections and their members, contact information for program officers, and helpful hints on the application process. Listed below are some NIH websites that are useful sources of information:

http://grants1.nih.gov/grants/index.cfm (overall NIH grants info site)

http://grants1.nih.gov/grants/forms.htm (electronic forms and instructions)

http://www.csr.nih.gov/studysec.htm (information on study sections, including rosters of members and review policies)

http://grants2.nih.gov/grants/funding/submissionschedule.htm (submission policies and deadlines)

http://www.niaid.nih.gov/ncn/grants/default.htm (National Institute of Allergy and Infectious Diseases site that contains helpful advice and tutorials, applicable to NIH grants in general)

Index